GLOBAL CARBON DISEASE!

DISPLACEMENT CHEMISTRY

A Double Mystery Whammy

WHAT HAPPENED IN 1950 THE YEAR OF MY BIRTH

Who and What put Earth on a Collision Course

with Extinction and Why

36% OF VOYAGE TO HELL TRAVELED!

"AS ABOVE,

SO, BELOW"

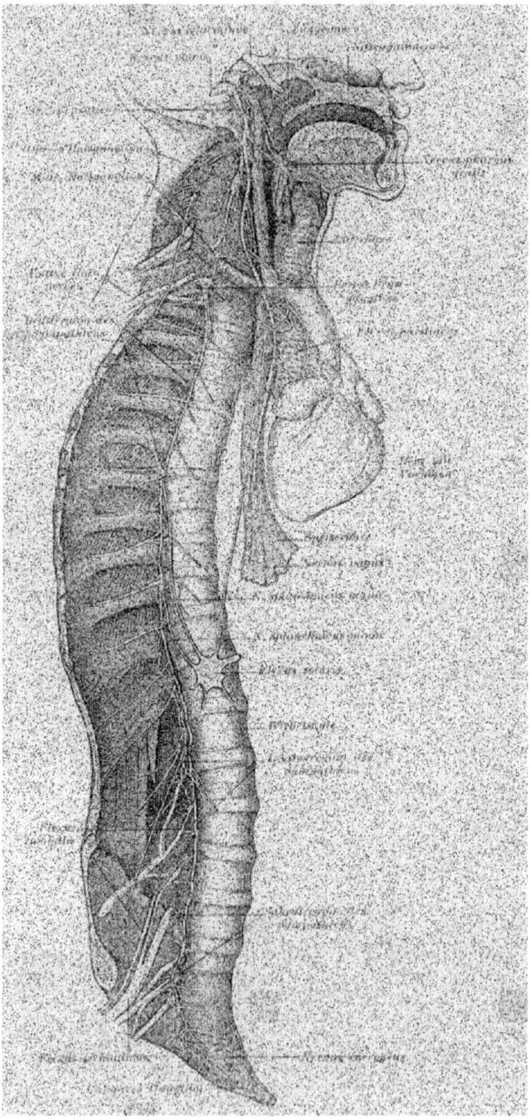

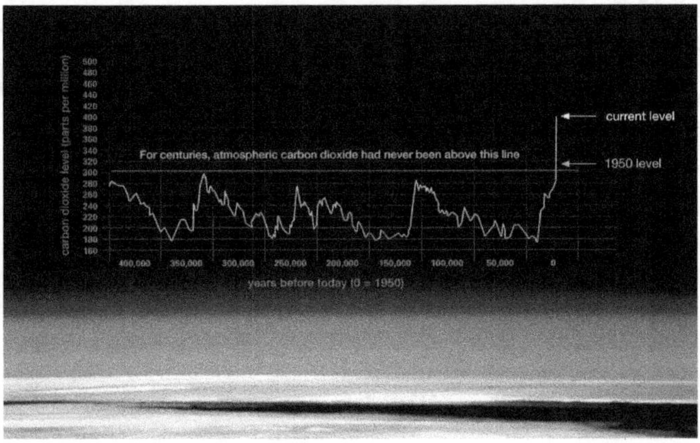

Are there people, or alien creatures, on Earth who would/could Geo-engineer our life support system for medical profits and moon travel madness?

The Germanic flavored 'New World Order' economic system lives by a balance sheet!

Can world business controllers depend on people getting ill normally in a healthy world?

Can world 'ultra-state' business managers depend on natural disease, or accidental illness of mind and body, for system profit, to happen naturally through wear or accident, at the right time for dividends?

Why not make the customers sick in a steady cash flow basis! Space ships kill with aluminum exhaust, so what, bring me another medal.

They are, We have!

'Cor·po·'rat'·tism' or corporatism is "a system of running a state using the power of organizations like businesses and trade unions that act, or purport to act, for large numbers of individuals". Most corporatists exhibit symptoms of common sense deprivation.

INTERNATIONAL ACADEMY OF CONSCIENCE OF LONDON

"The use of carbon dioxide shows a similarity between the phenomenon of the conscious out-of-

body projections and the experience of biological death. When carbon dioxide is inhaled in its pure form, it produces the death of the human body; inhaled in a small per volume amount (30%), predisposes the projection of the consciousness through the 'psychosoma' (OBE).”

What the frack!!!!! Yeah, take a breath relax and think about this before going farther

One Answer

Here

Is this the Answer to our current planet killing path?

Do we suffer from

Planetary Psychosis

Have I discovered a scientific reality that should change the way we THINK?

About living on Earth?

Or not!

Mind Altering Chemical

Evidence Is in Us

Air, Water, food and Hair

HAIR IT IS!

THIS IS IMPORTANT

FOR OUR SURVIVAL

SINCE 1950 WHEN I WAS BORN I AM BREATHING MORE THAN 33% BY VOLUME OF EXCESS MAN RELEASED CO_2. YOUR BODY IS NOT ABLE TO PROCESS AND EXPELL THIS EXCESS CO_2,

JUST LIKE THE MARCH TO ACIDITY OF THE MACRO-OCEANS OF THE PLANET.

WE ARE MICRO MINI OCEANS. WE ARE MOVING DANGEROUSLY CLOSE TO ACIDIFYING MAMMALS

AND ALL CELLS SUFFER AS CO_2 BECOMES

CARBONIC ACID AND SOLIDIFY AS CO_2 DEPOSITS IN OUR SPINE AND JOINTS AND BRAINS WHERE FLUORIDE ENHANCES THE IN THE BLOOD AND TISSES.

CO_2 MAY BE A MAJOR FACTOR IN ALL DISEASE FORMATION

THE BIG LIE OR DIS-UNDERSTANDING IS THE RISE IN CO_2 IS 415.52 PPM AND RISING TODAY IN MAY 2019.

CO_2 IN MY FIRST BREATH ON FEBRUARY 27TH AT 12:30 AM 1950 CONTAINED 280PPM OF CO_2 AND AT LEAST 1% MORE OXYGEN IN THE AIR MIX THAT GAVE ME LIFE.

If I were in a room, in June 2019, near sea level, breathing ambient normal un-balanced air, I am breathing 32% plus, and rising

daily, percent more co2 than

1950 and only 37% more

than 1750. Yeah happened

in my lifetime and I must have

caused It!

The true deep breathing yogi, by holding a breath a long time, builds co2 levels in their bodies and brain.

At 30% co2 Nirvana is possible and out of body and out of mind experiences are possible.

Mass killing can come from high CO_2 and nitrous oxide gases that push weak links over the edge.

NOW THE MYSTERY SOLVING

Walter Cronkite

World Media Icon

A seed thought was given me by Walter during a 1987 interview, he let me do for some reason, for our first *Survival of the Seas Society* documentary:

'Narragansett Bay: Never a Sabbath'

It was acclaimed on national television and media by film Narrator and Author Peter Benchley, as the *"most comprehensive film on the oceans ever, and I doubt it will* be topped ever. *He said it all"*

It was my first independent film while still a commercial fishing captain, or a trip mate, with captain friends in two fisheries. This was after a year on un-paid sabbatical from

full time captains' work, with CNN as field producer/creator of the first ever news documentary of the true state of the coast of America.

This film was the first time I really publically connected the dots. I connected all the dots of a coming storm, not so perfect, of revelation like environmental changes that were gaining speed

In 1979 I delivered, the first wave of evidence gathered OJT on deep water fishing vessels, with audacity, at a chance meeting on a dock in Rhode Island in 1979, with Ted Turner,

that meeting and my fine follow through, put my ideas on world television. Decades of information fragmentation required decades of search and seeing between the lies and the lines.

Until recently, I did not see, in order, enough of the pieces clearly of seeming un-writable books.

I do know, from study of vast historical evidence, that psychologically planned information control and fragmentation is paramount to dominion over minds.

This has kept non-hermetic, or dialectic logic thinkers, from ever understanding these planets changing happenings.

I was in a mental and emotional hibernation for 7 years since my last major film work.

My body was injured by a life of astronomical work at times. Some idiotic shaman wrote once suffering is a good thing.

It is thought that without heartache there would be no music. I agree in part. Joy brings great surges of music and lyrics from this soul. Yes, my mind wanders. Back to the mystery solved,

I required two spinal surgery in 10 years,

and typical of me, waiting to the last minute to remove fatal, or paralytic, deposits of chemically condensed co2. Without intervention from excellent surgeons in India and Arkansas. Both are very busy removing these from the young and the old.

Erectile Dis-Function

The first symptom of co2 deposits in men with spinal compression is erectile dis-function. Yeah more good news for the world bad for you girls. I never lost the flag pole completely. But size had reduced over time as my co2 deposit enlarged slowly with every breath of excess co2.

Attributing this to aging I used what I had,

Plus, a few tricks I had learned that always satisfies. I never in my life left a woman frustrated and dis-satisfied. But CO2 made me work harder to overcome this life changing situation.

While in Amsterdam, during November and December 2009, in a secret cannabis resin clinic (squat house), where I had reduced a tumor around my spine using THC concentrates IN THREE WEEKS. And gained confidence from the cannabis resin medicine. I was no longer concerned about the outcome of my GLOBAL CARBON DISEASE surgery.

Returning to Amsterdam from India, 6 weeks after my surgery, I fell into an exhaustion sleep, in a very cold room on a mattress covered heavy but comfortable. I had put a THC patch over my neck surgery and the pain went away as I slept. I forgot being told not to leave it on for too long.

The difficult recovery trauma and some 40 hours of flights between Mumbai and Amsterdam caught up with me.

After flying from India, I was stuck in Bahrain and London airports for many

hours of the travel sitting up after the surgery was a difficult time. But the moments when I remember feeling I had turned the healing corner was after three unbelievable wet-dreams.

I had not had one for a long time. The first was like an elevator coming from a deep place. I reached into the dark and grabbed a t-shirt, cleaned myself and fell, even deeper into sleep. In the darkened room, I came again, reached for my shirt cleaned again then more sleep.

The third time was so empting that I did not try to stop it. As I fell back asleep again I realized the spinal swelling was gone and my spinal cord was no longer compressed, was gone I was back to where I was as a much younger man.

The penial nerves had opened and I was back. I knew I was healed, for a while. Thank you, Sam, the mad like resin man of

Amsterdam.

Near death and the PTSD that follows and affects, at times haunts, in subtle ways like putting off the next surgery that is past due, again. Just like we all do on Earth, escape a catastrophe somewhere else, and it is business as usual. PTSD enhances the reality of just so much time to work with. "If not now than when".

Losing my nerve, at tired times, to this writing, with time running out, while confident that, with so much unfinished work and living the universe would never close the show before the final acts are seen.

An old man in Cuba told me after I ask him about still supporting the failing Revolution, he said, 'I paid the ticket to this show I intend to see the ending".

I know me well and have experienced the doldrums of learning and waiting for the

right time. It is now, that is a good thing. Like now for the last five months, when I starting writing again and finally found the music that let more out.

That was the critical missing element in my spiritual and activism chemistry.

When the time comes to get something out; when the magnetism of the need and the urgency takes hold; when you are caught in a just and needed orbit of dedicated ones, like me, we are helpless, but empowered by this force called evolution.

Evolution requires attention given to the reality that needs changing.

Through almost daily chance information, that gave me key pieces of the puzzle that brought clarity and it, is still happening to me. There seems to be a desperation in the collective conscience caused by the senses sensing we are chemically **out of balance.**

Time is pushing hard. I could not wait any longer to finish these decades of information and writing, but put away until now.

I have 7 books in circulation, final show-down writing, and assembling less recent works, began 24 December 2018, yes and more coming.

Balance: A fine woman, a Ukraine refugee, family all dead in Donbass War 12 thousand miles away, now in Russia gave me the damn breaker.

An insight I needed to stoke a fire in me out of smoldering resentment at stupid or evil planet killing cowardly controllers.

Another person's story and the emotions that came began empting the contents of a reservoir of a life, of vital thought and logical presentation, of learning important things.

Do you understand, I am helpless when it comes to being a teacher! A damn broke opening a scar reveling the cause of a life of expecting another, “shoe to fall”.

Continue reading and be as alarmed, into some boundary stretching action, as I am doing now.

Tonight, and every night and day for months getting this done ruled every cell. The damn broke and when the flow has flowed, I may be at some special kind of peace. For a second maybe, ha.

This is so real

Years ago, I had a moment when, I realized life was full of unseen turns that may bring joy or wisdom. Most often mutually exclusive turns.

That was the day I said to myself, and got no argument, “bring it on”. Anticipating that

'what' is next "Far Out" feeling of happening in waves. These happenings brought emerging understanding and the freedom that brought.

'COASTLINE CRISIS'

CNN 1984-84

I was an "enigma", said Peter after our first Port Aransas video shoot in 1984.

I had to look the word up first then thought, Yeah. I was an environmentally

alarmed by the decline I saw in the two oceans, The Gulf of Mexico and Atlantic

where I worked as a fisherman and captain. In time, I became an activist/producer/director/writer.

Somehow I bluffed my way into producing and directing television. It was my weapon of choice. 'Never a Sabbath' first aired on Rhode Island PBS and eventually world broadcast. I landed in front of Senate committees. What dummies!

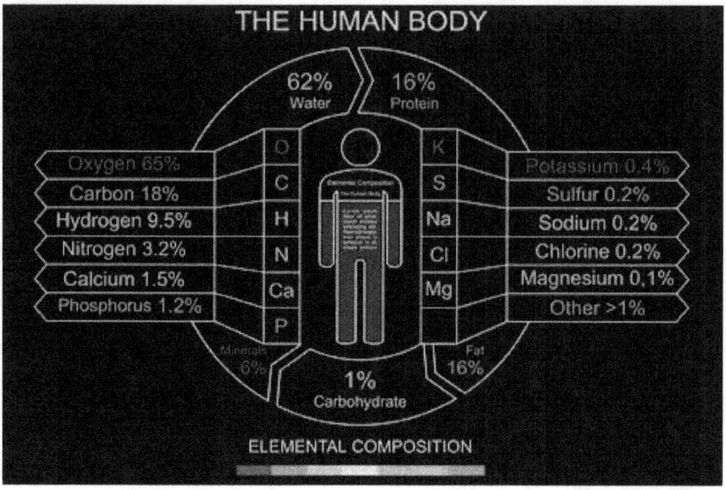

During the interview Walter told us something extra profound! I wondered for decades what the following meant and finally the reality was revealed to me to reveal to you. Action now or we may forever wish we had listened

Walter: "The entire matter of our environment air and water is something that is absolutely critical. It dumbfounds me that the public have not been pouring out into the street saying dam it I want take it anymore. It is that kind of a crisis but we still don't seem to awaken and it going to take some

kind of major disaster but the disaster may come too late"

He went on to relate to us an interview he did with a Nobel chemist who told him "humanity may put a combination of mind and rational behavior altering chemicals into the air, food and water and they may not be able to think themselves out of the future troubles (environmental)" Walter told me that the scientist paused, then said "and they may already have".

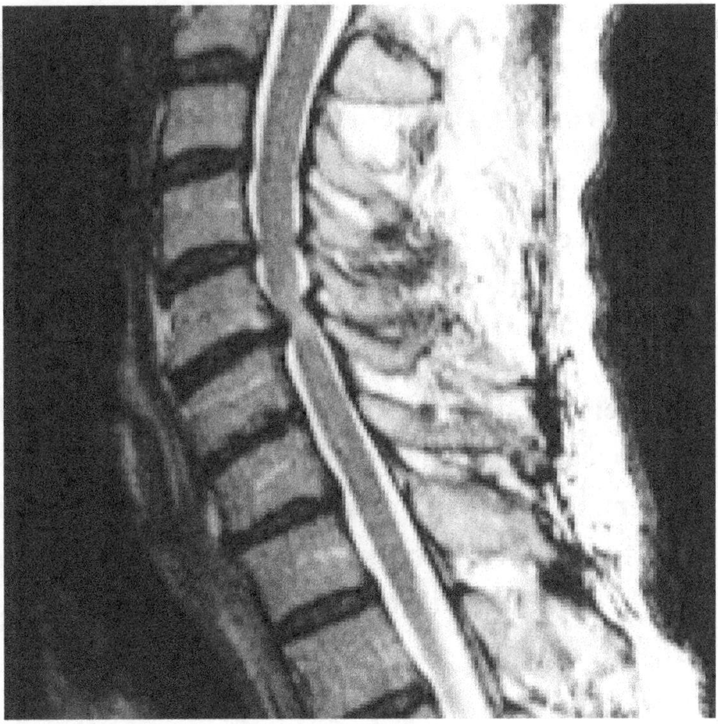

CO_2 as concentrated

Carbon Deposits in My Spine Almost Killed Me, TWICE

Dr. Martine Carroll

A Tribute from a Friend

2018

I had not seen Captain Garry Burris for 3 years when he entered my office in Hot Springs Arkansas, unexpectedly as usual, when coming home from the sea or from some activist mission.

Within minutes Captain Garry, looked me straight in the eyes and said very calmly: you are suffering from "Chemtrails" poisoning.

I knew this captain was a fine researcher and thinker and doer for almost 40 years. I listened to him because having observed his career, as a fisherman then an environmental activist, teacher and filmmaker I knew he was telling me a strong truth about something I did not understand, yet. I trusted him.

My first reaction, like almost everyone confronted with something so big, was show me "Chemtrails". I was not a blind believer because gathering accurate information has been a big part of my professional life.

When Captain Garry was a crewman on shrimping boats in the Gulf of Mexico and Campeche he would come home telling stories of destruction of the ocean. And his information and stories grew even more so when he became a fishing captain, then he would come home from the sea with new information and new stories from several oceans and finally from around the world.

Eventually he captained a tall ship for many years exploring the world's oceans in search of the truth.

He would then share observations not just concerning the state of the ocean he soon came to understand a greater story of the state of our world. As he would call it "our shared life support system".

He made predications about the future of our world and the sea. I wish he had been wrong but everything he predicted to me and to a writer at the

Arkansas Gazette in 1983 happened and is happening now.

That article was written by Margaret Arnold. Everything she wrote about his ideas and future vision came true. This 'chemo-madness' had to be true and to my dismay: It is!

To prove how organized the spraying program was Captain Garry, drove me up to the top of North Mountain in Hot Springs, Arkansas then set up his camera in front a little mountain pavilion. He said they would fly by in 3 minutes and they did. He had learned that they had a schedule and route to spray over.

While he was videoing me with his camera, in three minutes' spraying planes appeared over my shoulder. Later, I would come to understand through scientific evidence that the chemical trails were made up of the deadly combination of aluminum, barium and cadmium that came from a chemical in jet engine fuel and from nozzles along the tail and wings.

These long lasting, often swirling lines, became clouds and whited the sky for days at a time. One other MAD (Mutually Assured Destruction and

Mutually Assured dividends "sci-fi-elite" idea was to spray Nano-particulate coal ash. This idea shut down all the vital energy our bodies must have to make the vital vitamin d for us, mammals and even the crazies who did it.

They possess the knowledge and the antidotes but light energy or life energy is needed by all. They stopped that quick.

To show me more evidence Captain Gary asked me to let him cut several strands of hair from my head for a hair sample to send to a physician and lab in Louisiana.

Being in the health-related professional, and a clinical psychologist, with many years of experience working with Alzheimer's and dementia I was intrigued! Could he see Alzheimer's in me and help cure me?

He assured me that if the results of the hair test were negative then he would pay the cost.

Imagine the shock, my shock, at the results, a week later, that showed 12x's the expected range of toxicity in Aluminum in the first hair sample.

Most people have heard of reducing exposure from pots and pans, deodorant and any other contact to the mineral aluminum. Were those, maybe disinformation excuses, fed to the public to help hide this chemical spraying reality? Maybe?

After this it became easy for me to make the leap from low level crop duster over Arkansas and high level spraying for other reasons. I heard the secret program was called 'HAAD' or high altitude atmospheric delivery. What and why became questions needing answering!

A year later after going through extensive chelation with over 20 intravenous EDTA (Ethylenediaminetetraacetic Acid) chelation and followed with a pill capsule form of chelation of EDTA, a second hair sample was administered. My toxicity had dropped to 5% of what had been in my brain and body before chelation. I was recovering my former health and back to work.

Following this treatment, a mass of material was found in my bladder the size of a small baseball. Logically this was the concentrated material that the chelation drove from my body but a prostrate problem may have blocked the material allowing it to clump together.

After the surgery, I ask to see the mass and they refused to let me see my property or take it home. Taking home gall stones has been normal procedure for a long time and now I am told a government official came and got the mass. Maybe to stop me from putting 2 and 2 together?

Most of my physicians were critical of this hair sample and looked at me as though this evidence was just a fluke of exposure and that blood sampling could also reveal what the hair sampling had revealed. They were taught a wrong assumption to prevent real medicine from seeing the dangers of living in America and possibly planned and chemically administrated human obsolescence.

The liability issue looms for those found guilty of knowingly spraying this toxin on all of us. We are all in danger without the antidotes.

However, my thinking and mental processes had now vastly improved as well as other physical ailments. Aluminum poisoning takes all kinds of avenues in producing its physical symptoms. My best thanks to the universal providence that came with this physical dilemma. Here was a chance, with a clear mind again, to give a big thanks to Gary and his work to heal our planet and knowing that he had also saved my life twice gives me a greater sense of respect for this real old soul. He is truly a captain of many things including himself. Of

note the only physician I know of who regularly uses hair sampling is a veterinarian.

How to Defend your Selves

Hair sampling can be the best way to start investigating your levels of toxic metal exposure and disease caused by essential mineral depletion. Almost all dis-ease can be tracked back to disease fighting mineral depletion. Since most of my time today has been inside an office and differences exist between people, caution is the best part of valor.

Hair sampling can be the best method to begin your journey of finding out personally what your health future may be so if you are searching for answers this is the real deal.

Metal toxicity, in part, can result from years of vaccines, mercury dental fillings, food products additives, etc. over all nations of the world. But I had not had a vaccine in more than half a century.

It is simple deduction: In goes molecularly heavy toxic metals and out goes the essentials. This is "displacement chemistry" very simple stuff.

'Chemtrails', and contrails, contain plenty of toxic metals and other gaseous toxins that have been

sprayed and jetted upon you for many years. All the other causes mentioned above are not even a one percent of the physical impact of this spraying program.

If you do not believe the spraying truth try this: The reality of the massive Geo-engineered world poisoning atomic bomb testing that negatively altered the planet is not taught in schools (got to wonder why) but every inch of the planet is contaminated with radioactive bomb fall out material that, some or most, are still as radioactive today with half-lives, or decay lives, of millions of years, as it was the years it rained and snowed and slowly drifted onto a population of deafened and dumbed by the big lies. Lies from the controllers are coming from purposely inbred people now infiltrated into power (as Eisenhower warned) with great influence over Presidents and Generals.

Elections are a rig-able formality in a fooled democracy.

Today you might hear this spraying described as "Geoengineering". It is another 'MAD' program from the "scientific elite" that President Eisenhower warned America about this in 1961.

This destructive program has been going on since before the advent of the atomic bombs test in 1945 and the subsequent real reason the ozone and the sky were fried is the atom bomb use to kill and thousands of mindless tests.

They have told the big lie all along to cover for the sci-fi-elite bomb designers.

These high-altitude spray planes have operated unnoticed for many years but now exposure is imminent.

The weather modification aspect, claimed by the Air Force on their web site to be a "force multiplier", appear in battle field footage from over the Battle of the Bulge in 1945. Weather was used as a weapon and by clearing the clouds that battle went in the favor of the side with the most money.

Early Evidence

My skin color and slow thinking were all indications that my mind was no longer my best friend. My, mind, when clear, in order and fully active, was my tool to make a living: it was fading into no order. But now evidence of this spraying was in my brain: it was aluminum and other brain toxins. Still is but in low amounts. Body cleaning must go on after the enemy is discovered.

Before Garry came to me I did not know what had taken over my body: That I had been invaded. But with treatment the enemy had been subdued and my mind returned to near full power. Aluminum and other sprayed and absorbed metals are like some invisible nearing fatal enemy attacking all of us in a subtle way.

It is usually a function of good body and diet care and pure water if you are not losing your mind to aluminum, barium and cadmium and 65,000 plus chemicals in the mix we call our air supply; our life support system. But even with the best diet and air quality you will no doubt have these toxins in you and you will be assailed by all kind of disinformation about the sources. An example

"aluminum comes from cooking pans".

You could eat your entire cookware set and not have the levels I had in my body prior to my treatments. And now "they" are putting aluminum into our water supply as if fluoride and bleach is not enough of an assault. Dentist who recommend fluoride toothpaste should be better educated.

Captain Garry had been a longtime friend since 1975. We founded Survival of the Sea Society with others in 1980 after he convinced me, with strong evidence, to the need of an organization upstream of the oceans to fight for the rights of those downstream: He told me this was a fight for the living ocean and life on this planet and for the far nicer than human creatures whom he shared many wondrous intelligent moments when he was a fisherman on several oceans.

33 years ago, we traveled together that summer of 1985 (in his 240Z) and spoke over numerous radio and television stations from Arkansas to Atlanta to Maine and every state in between talking rationally about mans' pollution and over fishing affecting fishing in the Gulf of Mexico and Atlantic and all oceans via currents.

Like dirty undisciplined children we, the mass of humanity is without thought of future. Without thought, with few teachers teaching, we pour a deadly poisoning mix of oil, chemicals and trash into all our waters affecting our food supply and ocean oxygen production every second of every minute.

The good news is this: due to the decades of work Captain Garry and others did, we the people now have the knowledge of cause and the antidotes for positive health effects to at least lesson the planet sickness we are all involved in now.

I began to understand those strange clouds were more than vapor condensation trails from jets in the stratosphere that disappear quickly.

Frankly, not much of my attention had been paid to the skies even though being an outdoorsman with plenty of outdoors experience like canoeing, mountain bike riding and camping left me with plenty of time to observe these man-made clouds that control our now out of control weather. I saw but I did not see!

To my surprise many experienced pilots look up when not flying and see nothing unusual.

Piloting my own private airplane for over ten years plus extensive knowledge of clouds earned me enough knowledge to fly instrument rated. How did he know the timing of those spray planes at that precise time and what was the real purpose?

My guess was that a very experience sea captain, living by his wits in the open seas, will see, and pay attention, to all kinds of atmospheric conditions.

Now may be the time to find answers to questions concerning the cause of many of those symptoms, like brain fog on days the sky is white-out and not by clouds, we are experiencing.

These will be left unanswered until you take that first step get a hair sample and take the evidence to your local governments. Communities all over the United States have taken up this issue and are beginning to stop something that you never asked for or ever wanted to affect your children and your family.

And by the way, chelating metals out of your system needs to continue for the rest of your life if you see the need to correct this portion of your life. That metallic taste in your mouth may mean mental illness that I now understand may come raining and

floating down from the sky as toxic metals know to displace natural lithium a brain stabilizer mineral.

Remember living your life to its fullest means making a choice of common sense. No one else will hold your hand for making good choices.

My final thought to you the reader: Captain Garry is probably the greatest universal thinker of this century and maybe any other. Only a true dedicated tenacious genius could have, as a self-disciplined activist, accomplish a body of work essential for the awakening of mankind.

****

There is good in all bad, bad in all good

Positive can be negative and negative can be positive. That is the thought twister of now. Sell you a gas hog car or truck and convince you it is good while the exhaust from it never blows away. It deposits ***quickly*** in our bodies with a lethal forecast for all.

Words are the illusion makers and right words the

illusion breaker. This expose of correct thought can be the equalizer between those who rule and those ruled.

Almost all are carbon "victims". Guilty of our universal deceit of our selves. This information should be the game changer and the test of the

power structure of Earth. Yes and no, this is no test; this is the real deal; ship sinking; start pumping.

THIS INFORMATION REQUIRES CONTROLLERS TO GIVE A SHOW OF ABILITY TO RESPOND TO REALITY AND TELL THE TRUTH FINALLY.

A PERSON DYING IS MOST OFTEN TOLD AND GIVEN ODDS OF SUCCESS. SOME ARE TOLD OF ALTERNATIVES TO DEATH

We are equally toxic with our collective waste. Instead of action we get reaction from our rulers. We get an America filling with "Black Shirts". This is very reminiscing of 1930's Europe in Italy and German.

And what if these citizens shooting back at these men in black are the real defenders of democracy.

Then what are **YOU** going to do.

"Action through awareness, awareness through education"

I hope that this message of universality hits a cord in you. This book will empower big time, if knowledge is what you seek. Just think of the **power** of information, clear and concise, that we are in trouble that enable rational re-actions.

No more time to discuss ideology this concerns Darwin's thought of the survival of the fittest; it is correct. But here-in is the real deal to be afraid of or glad for the cover.

I am watching a PBS film about the militarization within America. The controllers are preparing for riots and total meltdown of our social and business infrastructure because, in part, of the information within this book.

I have publically, on radio, said that I am beginning to understand, at a personal level, the infiltration and the take-over of our world system into a concentrated few hands.

I am rethinking my position on this. Here it is:

bring it on, bring on the troops fill our streets and use that power to slow down our monstrously murderous destructive American system. It came and began with killing Indians and now IT IS culling us in a mass un-controlled and un-punished murder by co2/nitrous/methane emissions; by car and cow gases; by "chemo" planes cargo of sickness, toxic water AND suicidal like actions of rockets and the exhaust finishing the process of humanity culling. The genetic modification of plants is a major weapon of body chemistry modifications.

Cannabis is made illegal in 1939

The planned chemical drugging of America and altering thinking patterns depended on cannabis not being used.

Cannabis has been on the planet for longer than we have been here and is symbiotic to us and us to it. We have cannabis receptors in our brains that utilize the benefits of this brain food.

The mind expanding, neuron feeding and mood altering in the positive effects of cannabis was chemically a natural good wall between us and them.

LSD cracked the façade for many and allowed spiritual growth. Cannabis is on the planet for spiritual grow enhancements.

Cannabis is becoming the biggest weapon of negative use. I see cloning poor qualities into plants that are capable of cloning behaviors negative in humans.

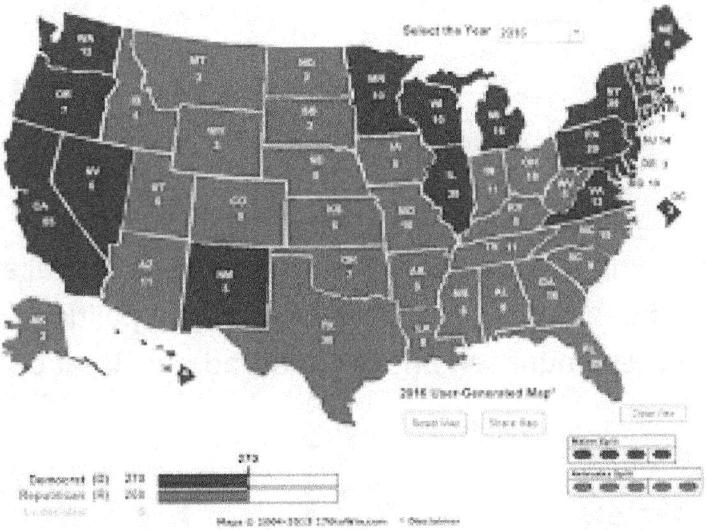

Red States-Black Uniforms

Bring it on please use these

organized police militarizing plans to control

America, to the bank or the food stores, completely, use it for good purpose.

If they are doing it, and **they are**, then do it right with the correct ideas of planetary environmental survivability; society's survivability.

Just do it:

ASAP

I am hoping this may help the world governmental system get it. These permeant controllers, and all political candidates, must see and understand the situation.

Therefore, we must **Slow**

Down.

This information can facilitate the recovery of Earth. I am doing *my part* by stretching my boundaries. How about you?

Those of us who have lived near the bone will be more prepared for the coming minutes than those who had the whole hog and let it rot to the bone, rather than change and distribute health instead of sickness

'Do Not Resist', a PBS documentary airing today July 9, 2019. Seeing the interview with the producer I was skeptical that he did not know that it was is a mastery of mass mind control programing. There are people called TV programmers they determine what information you receive and when you receive it.

It showed archive footage of former (sort of) Nazis in front of NASA with Senator Johnson.

They were the top scientist at NASA in charge of secret projects within America and with American money and without oversite from congress. I could hear the cheers of neo-cons who are as toxic as us chickens and they have been had by their controllers.

Stacking the deck maybe, those 1000-year business planners seem to think about most things. Except one: How to live in a co2 exterminated and aluminum poisoned human-insanity, plus all mammals

Germany had an agency in 1937 that claimed "to know more about every city and village in America than the people living there". **Yes,** they have been planning this mega-business deal for a long time. They conspired well to keep secret business plans and patents.

Germans may have killed the first Indians as an early step in this world dominion plan to **steal the world.**

Germans are monarchists. Get it King me, you subject to what I want.

Just now on PBS: The Boston Marathon bombing was called the beginning of the high-profile police invasion, after the "Black Shirt" infiltration, of America. This was for no good purpose but it intruded the fear factor. The "no-knock warrants protocol started in Key West 22 years ago.

There were 21000 such police home invasions in 13 states last year. These 13 states are 'red' states and wear black in uniform. Do the controllers know the eco-system of the planet is going to hell in a hand basket. Identical police tactics preceded the last culling in the last century. Chemicals in the water, of these confused police, causes as planned this behavior. This is programing and it works PBS, a twisting tool of the programmers is very guilty of sins of information omission. But they may be going as far as the rule of the game allows.

When it is 'allowed' to show a film about the take-over of America they are fulfilling a purpose. That purpose is to prepare us, and second; this gives plausible deniability to the infiltrators.

An example: Showing movies about super ships built for elite escape, when the world rolls over after the balance weight of Greenland melts. '*After*' the ships are built some high budget producer produces such a story. "The Day After Tomorrow', a fictional move about real things was another example.

I was asked by a commercial fisherman about that

movie and I told him that it is happening just at a different speed. He got it and thanked me for the clarity and here I go again.

Within is the information **purpose** to work together

We are doing this to us and this carbon disease may be nature's way to heal mother Earth of a cancer that is a yet awaken humanity

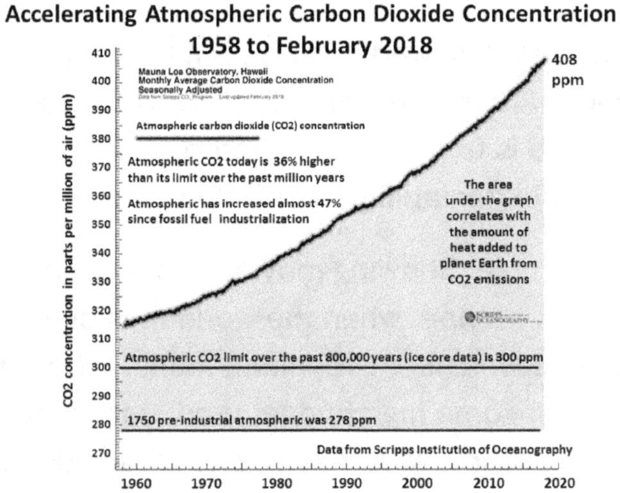

Today, July 9, 2019, co2 has risen to 415 ppm in one year. That level and rapidity of rise took millions of years and usually led to an extinction event.

'Darwins' are filled with awareness, and awareness of the territory is the first step in surviving in the jungle of life in a toxic world. Maybe, they may survive.

AtheistRepublic.com
Think of the rivers of blood spilled by all those generals and emperors so that in glory and in triumph they could become the 'Momentary' masters of a 'Fraction' of a 'Dot'.
Carl Sagan

IT IS NOT TO LATE

FEBRUARY 5, 1980

THE WORLD IS FALLING
DOWN
ASSUREDLY AND STEADILY
FEW MORE PAUSES
DOWN
ALL AROUND
I LOOKED FOR A PLACE TO HIDE
AND BELIEVE THIS BROTHER
IN ALL TOWERS
AND TO ALL SUPPOSED POWERS
IN ALL HOLES
CAN'T SEEM TO FIND

NO ONE WHO KNOWS

JUST WHAT TO DO

JUST KEEP THE FAITH

MY HEART KEEPS CRYING

KEEP DOING WHAT YOU ARE DOING

THAT WORKS

LETTING ONLY THE GOOD THINGS

MOVE YOU, MAKE YOURSELF BETTER

AND PASS IT ON

TAKE YOUR NEXT BREATH AND KNOW.

IT IS NOT TO LATE, BROTHER

During the process of acceptance

Of bad news

We go through physical and emotional shocks

In stages

Of lessening shocks

Denial, anger, depression, grief then acceptance

Are the stages we go through

All of us

As we try to wrap our minds around these issues

WE ARE IN DEEP SHIT. BUT WE

EVOLVED FROM DEEPER

SHIT. NOW IS THE TIME FOR THAT

LEAP

June 8, 2019

I add this new thought that evolved in me during the 39 years since writing that short story verse/poem.

Today, I finally understand that all the escalating destruction of our life-support system, since **1950** and the rapid change that is happening is fueling evolution.

To me evolution/creation is the reality. Evolution takes time on the physical plane to change mineral body forms.

But on the mental and DNA plane there can be a ***flashed*** imprint causing event trauma that polarizes the world in another direction.

Is this the tool and process of evolution? An example: A president declaring a "just war" to a divided nation in a "Pearl" event. Instantly the herd force of nature changed their collective direction like a flock of starling or a school of fish.

Hermetic knowledge of this cause and effect is a powerful tool used for centuries, in both directions, at the extremes of the pendulum swing.

But in test times like these evolution can happen in a flash at a speed much greater than the speed of light:

Thought Speed

Thought speed manifests in many ways. One I understand is the mind image of dropping a drop of water in an ocean and the distant reach of that ocean knows it without the passage of time. That is thought speed

June 10, 2019

Australian scientists concluded

"humanity will be extinct in **31**

years". One Military general agreed with this so, if correct, we are on the way out without

IMMEDIATE

Action

SLOW WORLD ENGINE DOWN TO HALF

THROTTLE AND MAKE

EVERY GALLON OR COAL VEIN

PAY DOUBLE

LET THE WORLD TAKE CARE OF ITS

PROBLEMS AND FOCUS ON AMERICA

NOW THAT IS PATRIOTISM

My question: If we are all gone in

31 years, thank god sort of, how

will the HUMAN die offs take place and is it going on and only I see the ***mechanism of extinction***.

Keep reading if you want answers so you can prepare your spirit for this **POTENTIALLY 'Armageddonish'** future.

A first sane action would be for couples stop having multiple children. You will see them sickened and died before it is their time. USE EMBRIO AND SPERM BANKS

ACTION AT LIGHTENING SPEED

NOW

OR ELSE

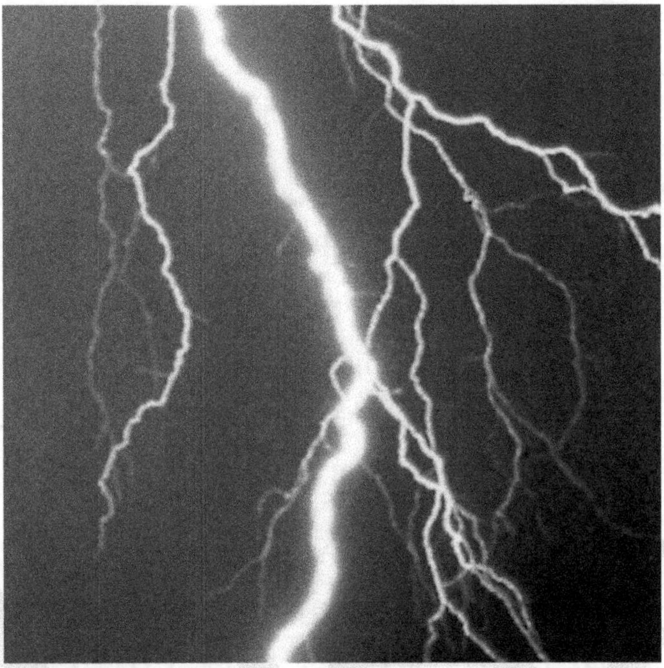

CRACKING THE CODE

The Big Problem

Too Much **Co2** and Reactive Halogens, In **US** PLUS nitrous oxide, or laughing gas, and some real mad creeps experimenting with your planet without “Full Disclosure or democratic permission”.

The NASA space program operated by Nazi planners of the New World Order is the culprit in total aluminum ion contamination

world-wide from space shuttle exhaust and chemo-plane spraying. Each shuttle flight releases **170 tons of AL and destroys 30 percent of local ozone** and NASA/Nazi rocketeers are the “destroyers” responsible for 10 percent of the loss of planet shield total.

Russia, apparently more caring of the life-support system, changed it fuel formula and not destroys 2000 times less ozone that NAZA: we won we bested Russia at something some of us would have preferred a loss with less pollution.

Today entrepreneurs are destroying more ozone with rocket test to develop space tourism. They do not give a damn. Evidence is a business formality and redaction works.

I will seek an immediate injunction against any new missile launch to stop farther ozone and magnetosphere destruction

MINDLESS PROCREATION

MINDLESS PROCREATION

IS THE SECOND.

Abortion is maybe a reaction to instinct saying the cattle pen is too FULL or women are too toxic to hear their bodies telling them they are ovulating and to use alternatives for their men to be normal.

I MUST ADMIT AN UNCOMFORTABLE FEELING TOWARD SOME WOMEN AND UNWANTED PREGNANCIES AND WOMEN NOT USING HAND AND MOUTH TO CONTROL UNWANTED HUMAN BABIES.

THE RISE IN SUICIDES MAYBE CAUSED BY CO_2, (co2 is a heavy gas it stays where it falls), NITRUS AND METHANE; MAYBE AN ACT OF DARWINISM IS SELF CONTRIBUTION PROBLEM ELIMINATING.

IF YOU LOVE YOUR FUTURE BABES KEEP THEM INSIDE FOR A WHILE TO SEE THE FUTURE UNFOLD.

CLINICS TO KILL BABIES AND CLINICS TO INPREGNATE THOSE TOO TOXIC TO CONCEIVE OF DNA DEFECTIVES WHO SHOULD NOT HAVE CHILDREN. ADOPT, MUMBAI STREETS FILLED WITH MOTHERLESS CHILDREN.

ASAP

"You have been alive forever to be here now"

We have survived events; we are living proof of Darwin's survival of the fittest. So, continue reading, get informed, get a

'Darwin Edge'. Then act! Get prepared!

Some Answers

Before the next level of the story: There is hope and **BODY** antidotes to our toxic world SITUATION: Chelation, constant mineralization, cannabis tinctures, chemical free water, carbon/co2 'hyperbaric de-toxic', aggressive alkalization of whole body

and **Darwinism**

This information may save your life. It certainly will improve your odds. If you have gotten this far you are a Darwinist.

You have the inquisitive gene and seek answers to riddles. Welcome hail Darwin. Is this a bit larkish, yes? But the fit who survive do things differently

HOW DID

A FISHERMAN/POET/ACTIVIST ASK, AND ANSWER VITAL, BEFORE UNASKED QUESTIONS ABOUT SURVIVAL OF THE FITTEST?

My answer: EASY! IT CAME WITH THE TERRITORY OF BEING AN INVOLUNTARY CAPTAIN, OF MANY A MORAL VOYAGE, ATTEMPTING TO HELP STEAR EARTH AWAY FROM THE SHOALS OF STUPIDITY. AND IT TOOK DECADES OF THOUGHT AND COLLATING.

SOME CALL ME INQUISITIVE OTHERS CALL ME NOSY. HERE IS ANOTHER ANSWER, I CARE FOR THE CARELESS REGARDLESS.

'I AM THE IMPROBABLE, DOING THE IMPOSSIBLE WITH REASONABLE RESULTS'

GOOD, YOU HAVE READ THIS FAR

(Did I get your Attention)

NOW, HERE IS THE MY LIFT BECAUSE NOW THE WEIGHT IS COMING OFF ME A LITTLE MORE, FINALLY, ITS BEEN A LONG TIME AND NOW IT IS OUT OF ME AND IN YOU.

NOW, YOU MIGHT NEED TO GROW UP AND FACE THE SITUATION OR ELSE BE BLINDED WITHOUT RECOVERY TIME.

THIS IS A BIG PROBLEM THAT SHOULD MAKE YOU EVOLVE QUICK, AWARENESS IS EVOLUTION ALSO. SOME WILL MELT DOWN AND THAT IS DARWINISM.

THERE IS A MAJOR PLANET ALTERING AND FUTURE SHORTENING HEALTH EFFECT CAUSED BY MAN INDUCED CO2 TOXICITY. THAT TOXIC MAY HAVE PASSED A TIPPING POINT, IN MY INTELLIGENT AND SCIENCE INFORMED VIEW, ON MAY 10, 2019. THAT DAY NOAA SCIENTIST WENT FRANTIC. ON THEIR WEBSITE ONE SCIENTIST WROTE,

"ARE YOU LISTENING YET!"

WWW. CO2.EARTH

THIS DAY LOWERED OUR ODDS DOWN CONCERNING HUMAN/MAMMAL SURVIVAL.

THERE IS A REAL THREAT TO SURVIVAL OF THIS LIFE PLANET.

OUR EVOLUTIONARY CREATION PURPOSE

WILL NOT BE FULFILLED

IF WE FAIL NOW

HERE IS THE NOT YET OBVIOUS: CO_2 HAS MIND ALTERING AFFECTS. CO_2 ALONG WITH NITROUS OXIDE (AKA LAUGHING GAS) ADD METHANE THAT LOWERS OXYGEN ABSORPTION AND YOU HAVE A PERFECT STORM OF MIND NUMBING CHEMICALS.

THROW IN THE IMPENDING HUMAN BLOOD ACIDIFICATION CORRESPONDENT TO THE ACIDIFICATION OF THE PLANET BLOOD, OR THE OCEANS, AND THE STORM BECOMES A HURRICANE.

A REEF DYING SOMEWHERE ELSE, IN AN ACID ENVIRONMENT, IS THE CHEMICAL MIRROR OF OUR HUMAN ACID KILLER STALKING US.

EARTH'S ENVIRONMENTAL DECLINE WILL END WITH THE NATURAL SELECTION

ACTION OF EXTINCTION.

THE EARTH AS A LIVING ORGANISM IS REACTING TO OUR EXCESSES.

THE QUESTION ON ALL THINKING MINDS SHOULD BE HOW TO SURVIVE AN EXTINCTION EVENT.

THERE ARE SURVIVORS OF PAST EXTINCTION EVENTS OBVIOUSLY, WE ARE STILL HERE.

HOW **DO WE SOLVE THIS PLANET KILLING SITUATION, EASY, LIKE EATING AN ELEPHANT, ONE BITE AT A TIME?**

SLOW AND RADICALLY REDUCE WHAT WE ARE DOING THAT IS CAUSING THIS DILEMMA. BY USING SOUND BUT DANGEROUS EVIDENCE TO FORCE AWAKEN, IN MASS DEMONTRATIONS, OUR WORLD LEADERS WITH THIS EVIDENCE AND ASK WHY

IS THERE LACK OF SANE ACTION BY LEADERSHIP?

HAVE OUR LEADERS PASSED A LINE OF MIND AND BODY TOXICITY THAT HAMPERS CLEAR THOUGHT? A CRONKITE MYSTERY CLUE?

YEARS AGO, I HAD EQUIPMENT TO TEST OXYGEN LEVELS IN LOCAL AIR SUPPLY AND TRAVELED AROUND THE COUNTRY GATHERING INFORMATION FOR THESE BOOKS.

21 TO 23 PERCENT OXYGEN IN THE AIR IS NORMAL MIND BALANCED GAS. IN FRONT OF THE WHITEHOUSE WE FOUND 18 PERCENT. IN MIAMI, EVEN LOWER.

IS IT ANY WONDER WHY NOTHING RATIONAL COMES FROM INSIDE THE WASHINGTON BELTWAY?

THE WASHINGTON POST REPORTED ON A SERIES OF STUDIES BETWEEN 1973 AND 2011 THAT SHOWED SPERM COUNT OF MEN FROM AMERICA, AUSTRALIA AND UK DROPPED 52.4 PERCENT. THIS DECLINE IN MALE FERTILITY

HAPPENED IN JUST 40 YEARS. DECLINE AMONG MEN IN AND AROUND THE DC BELTWAY WAS THE LOWEST. ***DC WATER IS HIGHLY CHEMICALLY TREATED WITH MIND ALTERING CHEMICALS THAT PENETRATE THE SKIN LIKE A SILENT LEGAL SUBTLE WEAPON.***

THAT IS SUCH GOOD NEWS

70% OF ISRAELIS ARE FLUORIDE TOXIC ALSO

THESE COUNTRIES PUT RAT POISON AKA FLUORIDE INTO THEIR DRINKING AND BATHING WATER. FLUORIDE PENTRATES THE SKIN.

FLUORIDE HAS A DIRECT AFFECT ON SPERM COUNT. ANOTHER TOOL OF KENNEDYS' WARNED ABOUT SECRET WAR.

A **GOOD ONE**. THE UNITED STATES IS THE MOST FLUORIDATED COUNTRY.

HERE IS SOME MORE OF THE EVIDENCE OF A CORRESPONDING ***CAUSE AND EFFECT*** RELATION TO MIND AFFECTING GASES IN OUR AIR SUPPLY/LIFE SUPPORT SYSTEM ON AMERICAN SOCIAL CONSCIENCE PROGRAMMING AND OUR SPIRALING

MORALITY DECLINE.

THIS MAY BE PRECURSOR TO THE KILLING IN SCHOOLS AND OTHER PLACES. IS THIS THE FIRST SIGN OF A COMING MASS MURDERING OF ONE ANOTHER WITHOUT CONSCIENCE BECAUSE OF THESE GASES THAT ARE RISING DAILY? **Yes!**

Daily!

CAN WE SURVIVAL THESE REVELATIONS?

YES, THE FITTEST

ASAP

AND THOUGHTFUL

WILL, EASILY, IF ONE THINKS IN BALANCE AND "TAKE NOTHING FOR GRANTED" AS EISENHOWER WARNED

LISTENING TO THIS BOOKS' MESSAGE AND ACTING FOR YOUR SELF INTEREST MAY BE EVOLUTION IN ACTION.

I ASSURE ALL THAT THERE ARE A FEW IN HIGH PLACES WHO ALREADY KNOW PART OF MY

EVIDENCE AND THESE BOOKS MAY EVEN THE INFORMATION SUPPRESSION PLAYING FIELD

ALLOWING CLEAR REASONED THOUGHTS

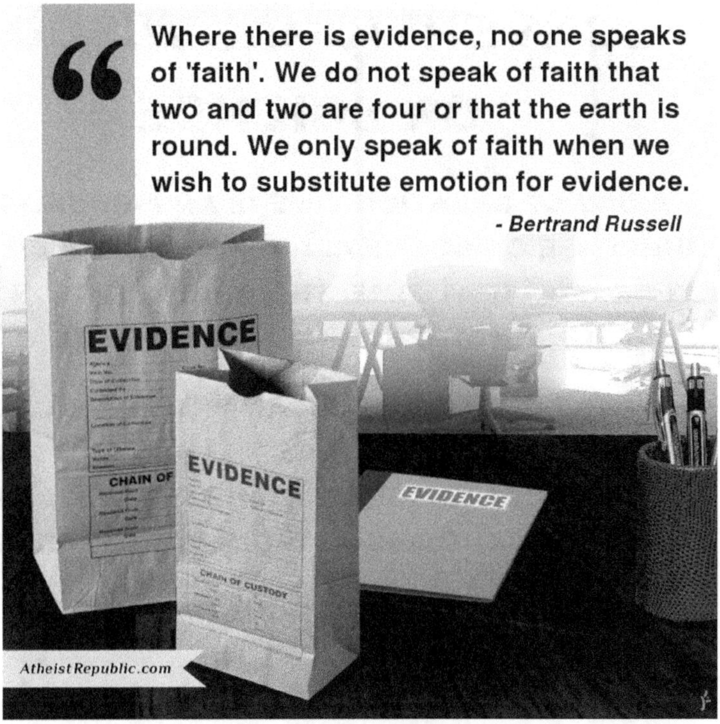
“
Where there is evidence, no one speaks of 'faith'. We do not speak of faith that two and two are four or that the earth is round. We only speak of faith when we wish to substitute emotion for evidence.
- Bertrand Russell
EVIDENCE
CHAIN OF
EVIDENCE
CHAIN OF CUSTODY
EVIDENCE
AtheistRepublic.com

NOT ALL

I HAVE BROUGHT FORTH IS KNOWN BY CONVENTIONAL WISDOM OR SCIENCE

A DAY OF CREATION TIME IS AN A-ROUND THE UNIVERSE CIRCLE OF APPROXIMATELY A BILLION YEARS OF LIGHT TIME. THIS WAS CREATION TIME OR THE 7 METAPHORICAL DAYS OF CREATION

THIS KIND OF PHYSICAL-METAPHYSICAL UNDERSTANDING GAVE ME THE KEY AND OPENED MY DOOR OF UNDERSTANDING EARTH SCIENCE IN THE MACRO AND MICRO SIMULTANEOUSLY

MOST ARE ORIGINAL THOUGHTS ARRIVED AT

WITH DIALECTIC LOGIC.

TO ME DIALECTIVE IS SEEING A CLEAR PICTURE EVEN IF THE INFORMATION IS UNCLEAR BY CONVENTIONAL WISDOM. IT SOLVES CONTRADICITONS

AFTER READING THIS YOU SHOULD THINK MORE CORRECTLY IN ACCORDANCE TO THE REALITY OF OUR TIMES AND THE CONTROL OF INFORMATION.

BETTER SAID WOULD BE: INFORMATION IS KEPT FRAGMENTED IN PATTERNS OF PLANNED MEDIA CHAOTIC CONFUSION.

THE CONTROLLERS ARE DEPENDING ON THE MIND NUMBING AND THOUGHT CONFUSION EFFECT OF A TOXIC AND CHEMICALLY DRUGGED WORLD TO PREVENT MASS AWAKENING AND MASS ACTION.

ALL OF US ARE WILLING VICTIMS OF OUR ENERGY OVER-CONSUMPTION MADNESS AND MINDLESS CO2 FLUSHING

INTO OUR OWN AIR SUPPLY

YOUR CAR TAIL PIPE

RELEASES CO2 GAS

INTO YOUR BODY

IMMEDIATELY AFTER RELEASE

INTO our LIFE-SUPPORT SYSTEM

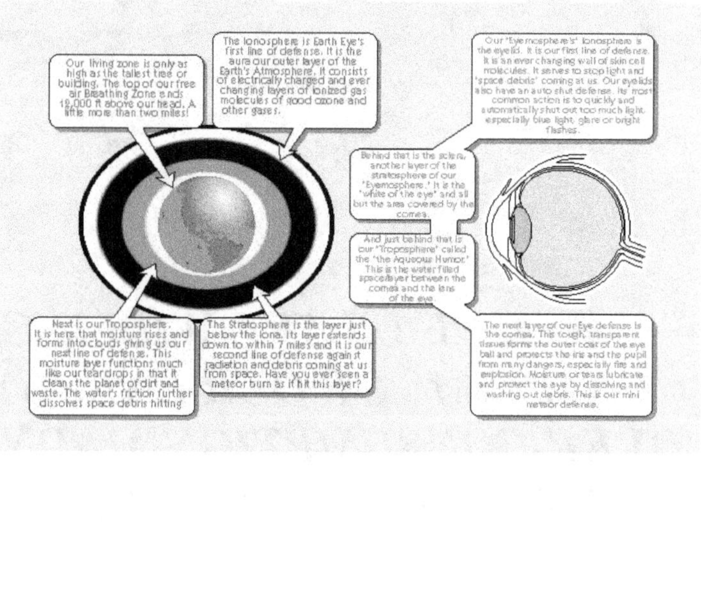

FAR FETCHED THOUGHT WE THINK

1986 Arkansas Farm

IT IS A FAR-FETCHED THOUGHT

WE THINK

WHEN WE THINK PEACE

OUT IN THE FUTURE

NO FOOD FOR YOUR CHILDREN

KILL YOUR NEXT-DOOR NEIGHBOR

FOR A BONE FOR A STEW

JUST A THOUGHT

OF FOOD AND DIRTY WATER

WALK PAST YOUR BEST FRIEND

OF YEARS, OF LIFE

ON A STREET

IN THE FUTURE

IN FEAR OF BEING ASK

FOR A BITE FOR A CHILD

OFF AN ABYSS

WE PLUNGE HUMANITY

VOID OF PRIDE

OF HUMILITY

OF FRIENDSHIP

PEACE AND LOVE

STOP, *THINKLOOKAROUND*

LISTEN

THE SOUND OF THE CROWDS ARE GROWING

THE PRICE OF LIVING IS A BURDEN EVEN NOW

TO NEAR ALL--- SAVE THE FEW.

IN A WORLD OF MANY

ALL THIS EVEN NOW

IN A WORLD OF GREEN CLEAN RIVERS

AND FLOWING FIELDS OF WHEAT

THAT AWE EVEN A SAILOR

OF VAST SEAS

IN A WORLD IN A UNIVERSE

THAT IS YOUTH TIME PERSONIFIED

MOTHER, MOTHER EARTH

YOU BORE THESE MANY CHILDREN

FOR WHAT ITS WORTH

AND IT IS LIKE LITTLE CHILDREN

BADGERING THIS MOTHER

WITH PITYING EXCUSES

FOR FUCKING, EVERYTHING DOWN

SOME EXCUSE, SUCH ABUSE

MOTHER, MOTHER EARTH

DAILY WE STEP CLOSER

TO THE END

EACH DAY WE

PUSH YOU

CLOG YOU

SMOG YOU

WE BREAK THE COMMANDMENT

TO HONOR YOU

ANOINT YOU WE DO

WITH OIL AND PESTICIDE

STEPS TO OUR OWN GENOCIDE

MOTHER EARTH,

(PLEASE, PLEASE TAKE - ADDED JULY 2019)

MY ONE-MAN APOLOGY

FOR MY PART FOR CLOGGING YOUR ARTERIES

PUSHING YOU CLOSER TO EARTH CORONARY....

IT IS A FAR-FETCHED THOUGHT

WE THINK

WHEN WE THINK FUTURE

P.s.

YET IF WE TRY

AND BELIEVE AS WE DO

EARTH'S LOVE OF HER CHILDREN

WILL BRING US THROUGH

(I CANNOT READ THIS WITHOUT TEARS EVEN AFTER ALL THE YEARS)

1950 Again

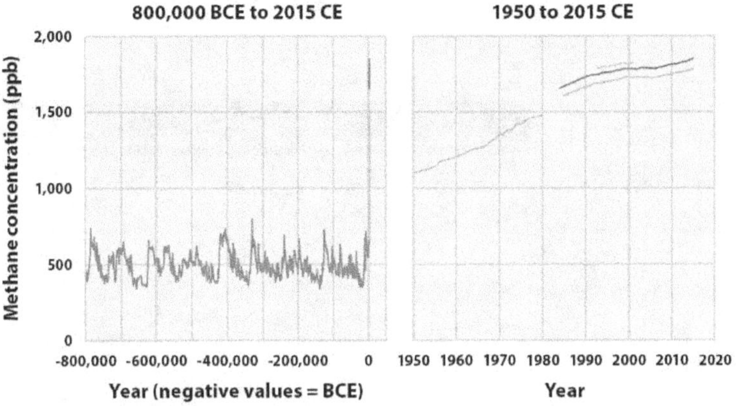

Starting to feel like the Snuffy Smith character with the dark cloud following *HIM WITH* ***THE RAIN***. Still,

in the quiet I surmised **a**

Humanity Unifying Factor

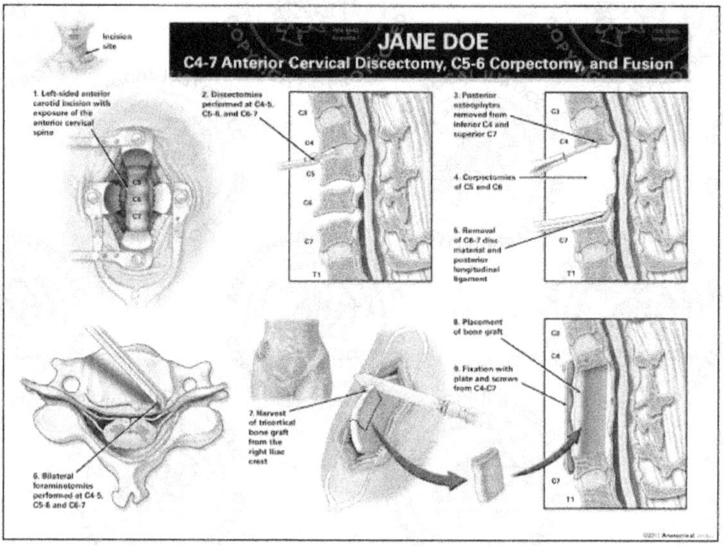

GLOBAL CARBON DISEASE

WE ALL HAVE THIS *Potentially*

FATAL

Within this generation

This is growing inside us; AN EVOLUTION catastrophe

IT TOOK A POET TO THINK

THIS ONE

INTO THE OPEN

MY PLANETARY THEORY/CONCLUSION OF CO_2/NITRUS/METHANE EFFECTS ON ALL LIFE! THAT EFFECT THE BRAIN.

CAN CO_2/NITROUS OXIDE/METHANE ALTER THE MIND OF MAN? NITROUS OXIDE, LAUGHING GAS IS RISING FASTER THAN CO_2, METHANE EVEN FASTER. THIS GAS COMES FROM CATTLE WASTE, AGRICULTURE AND OTHER MAN INDUCED SOURCES

HERE IS A GOOD QUESTION: HOW DID A COMMERCIAL FISHERMAN OUT

THINK THOUSANDS OF PAIDED AND UN-PAIDED THINKERS? THAT IS SIMPLE.

IT TOOK A POET AND HERE IS SOME

MORE EVIDENCE

While in college, in Arkansas, I was asked to take a new course in 'Oriental Thought'. My professor, Dr. Corcoran told me something important about debating an idea.

First, state the idea with total confidences. Second, defend it with your own logical interpretation of information then a summation with confidence and audacity, plus audacity and finally with audacity.

Corcoran told me I would pass his course, even if I was wrong logically, by the act of original thought woven with (dialectic) logic delivered with confidence. All wars are started with variations of this method or program. Is a life a series of programs? Here goes what I think will change the course of humanity if acted upon with confidence of change leading to timely action. My thesis of my knowledge and crystal clear information, started with a mystery and a hint and un-taught observation.

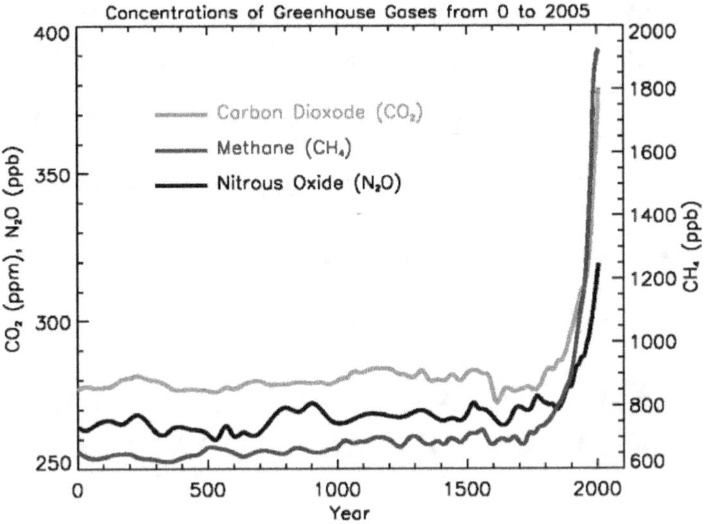
Concentrations of Greenhouse Gases from 0 to 2005
Carbon Dioxode (CO_2)
Methane (CH_4)
Nitrous Oxide (N_2O)
CO_2 (ppm), N_2O (ppb)
CH_4 (ppb)
Year
400
350
300
250
2000
1800
1600
1400
1200
1000
800
600
0
500
1000
1500
2000

Correspondence! Yes, as we burn fossil remains of compounds of carbon and as burned each molecule of carbon attaches to and permanently locks oxygen forming co2. No longer digestible by our lungs

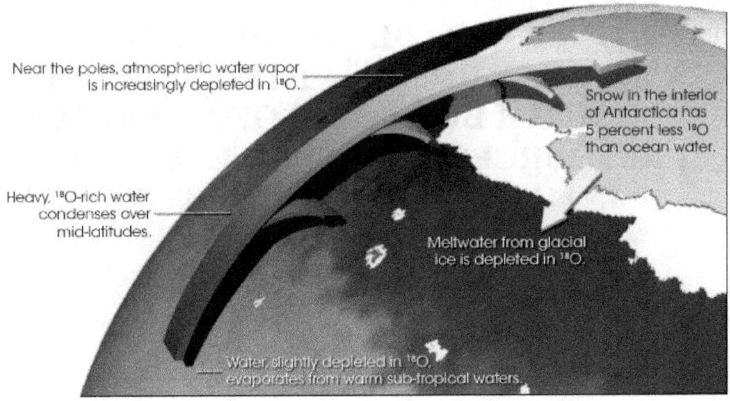
Near the poles, atmospheric water vapor is increasingly depleted in ^{18}O.
Snow in the interior of Antarctica has 5 percent less ^{18}O than ocean water.
Heavy, ^{18}O-rich water condenses over mid-latitudes.
Meltwater from glacial ice is depleted in ^{18}O.
Water, slightly depleted in ^{18}O, evaporates from warm sub-tropical waters.

WHAT ARE THE PRIMARY WEAPONS OF MASS MIND CONTROL TO MASK REALITY?

THE OBVIOUS BUT UN-TAUGHT

CAUSE OF ALMOST ALL DISEASE

A THREE PRONG STUPID ATTACK, MAYBE?

OR A CONSPIRACY OF SILENCE

METHANE DISPLACES OXYGEN AND KILLS BY LOWERING SLOWLY OXYGEN LUNG CAPACITY

ANOTHER IMPORTANT EXAMPLE OF DISPLACEMENT CHEMISTRY

NITROUS WAS USED BY THE NAZI'S IN EXPERIMENTS OF MIND ALTERATION. SOME OF THE RESEARCHERS BROUGHT TO AMERICA IN THE PROJECT PAPERCLIP PLAN, AND EARLY PREDECESSORS WERE THE ONES WHO DEVELOPED THE PLAN FOR AMERICAN DUMBMATION-MY WORD-AND SUPER BUT TOXIC OF AMERICAN RESOURCES WHICH RELEASED POISON INTO OUR LIFE-SUPPORT SYSTEM.

CO_2 REEKS HAVOC IN EVERY CELL

IN EVERY ORGAN SITTING THE STAGE FOR DISPLACEMENT CHEMISTRY

MANIFESTATIONS TO INVADE AND CHANGE THE NATURAL PROCESS OF AGING AND DYING INTO A PREDITABLE COMMODITY FOR SYSTEM CASH FLOW AND PROFIT

THE BODY SYSTEM DRIVERS ARE VITAMIN D AND IODINE

LACK THESE AND NOTHING WORKS PROPERLY

I now understand that this mix of mind altering chemicals is the answer to the riddle of Cronkite. This compounding mix, with predictable results, began as an experiment of omission, is now out of control.

1. Lindner, Kurt
2. Jungert, Wilhelm
3. Debus, Dr. Kurt
4. Fischel, Dr. Edward
5. Gruene, Dr. Hans F.
6. Mrazek, Dr. William
7.
8. Schlitt, Dr. Helmuth
9. Axter, Dr. Herbert
10. Vowe, Theodor K.
11. Beichel, Rudolf
12. Helm, Bruno K.
13. Holderer, Oscar
14. Minning, Rudolf
15. Friedrich, Dr. Hans
16. Haukohl, Guenther H.
17. Dhom, Friedrich
18. Tessmann, Bernhard
19. Heimburg, Karl L.
20. Geissler, Dr. Ernst
21. Duerr, Friedrich
22.
23. Milde, Hans W.
24. Luehrsen, Hannes
25. Patt, Kurt E.
26. Eisenhardt, Otto K.
27. Tschinkel, Dr. J.G.
28. Drawe, Gerhard P.
29. Heller, Gerhard
30. Boehm, Josef
31. Muehlner, Dr. J.W.
32. Rudolph, Dr. Arthur
33. Angele, Wilhelm
34. Ball, Erich K.
35. Heusinger, Bruno K.
36. Novak, Max E.
37. Mueller, Dr. Fritz
38. Finzel, Alfred J.
39. Fuhrmann, Herbert
40. Stuhlinger, Dr. Ernst
41. Guendel, Herbert
42. Fichtner, Hans
43. Hager, Dr. Karl
44. Kuers, Werner R.
45. Bergeler, Herbert
46. Maus, Hans H.
47. Schwidetzky, Dr. W.
48. Hoelker, Dr. Rudolf
49. Kaschig, Erich K.
50. Rosinski, Werner
51. Scharnowski, Heinz
52. Vandersee, Fritz
53. Urbanski, Arthur
54. Tiller, Werner
55. Woerdemann, Hugo
56. Schilling, Dr. Martin
57. Schuler, Albert E.
58. Lindenmayr, Hans J.
59. Zoike, Helmut
60. Paul, Hans G.
61. Rothe, Heinrich C.
62. Roth, Ludwig
63. Steinhoff, Dr. Ernst
64. Reisig, Gerhard H.
65. Klauss, Ernst K.
66. Weidner, Dr. Hermann
67. Lange, Hermann
68. Paetz, Robert
69. Merk, Helmut
70. Jacobi, Walter W.
71. Grau, Dieter E.
72. Schwarz, Friedrich
73. Von Braun, Dr. Wernher
74. Wittmann, Albin E.
75. Hoberg, Otto A.
76. Schulze, William A.
77.
78. Thiel, Dr. Adolf K.
79. Wiesman, Walter
80. Buchhold, Dr. Theodor
81. Rees, Dr. Eberhard H.
82. Hirschler, Otto
83. Poppel, Theodor A.
84. Kroll, Gustav A.
85. Voss, Werner E.
86. Beier, Anton
87. Zeiler, Albert
88. Schlidt, Rudolf H.
89. Steurer, Dr. Wolfgang
90. deBeek, Gerd W.
91. Millinger, Heinz
92. Dannenberg, Konrad K.
93. Palaoro, Hans R.
94. Neubert, Erich W.
95. Sieber, Dr. Werner
96. Hellebrand, Emil A.H.
97. Hosenthien, Hans H.
98. Bauschinger, Oscar
99. Michel, Dr. Joseph
100. Scheufelen, Claus
101. Burose, Walter
102. Fleischer, Karl
103. Gengelbach, Werner
104. Beduerftig, Hermann M.
105. Hintze, Guenther

Know thy enemy, Know thy self

The NAZI extension of power and idea implementation into America, and other places, began as Project Paperclip scientists with help founded NASA when brought to America and infiltrated into positions of control over the air water and use of food additives that are part of the dumbing down of America in progress. They were given dominion over space travel. At least half of the first 1947 cold war game scientist waves were sent to Russia. They returned in 1947 to Germany and many came to America as warriors in business suits and friend in high places.

Russia thought they had learned enough from them to play the arms race gain game.

They were chosen and financed by the Warburg German infiltrators who established the monetary control needed for dominion over America and the World

These systems invaders were brought in groups of implementers and invaded into companies of significant value to the world business plan until 1992. Yeah, in your face. Eye of Horus guys we see then represented on our paper air money.

It is all a planned game of war profit and now war on our bodies and children for medical cartel profit. Don't blame all doctors because few know this information.

Alas, the fate of the planet was sealed, before these scientists arrived in 1947, by the mad atom bomb dreamers and makers along with infantile military minds.

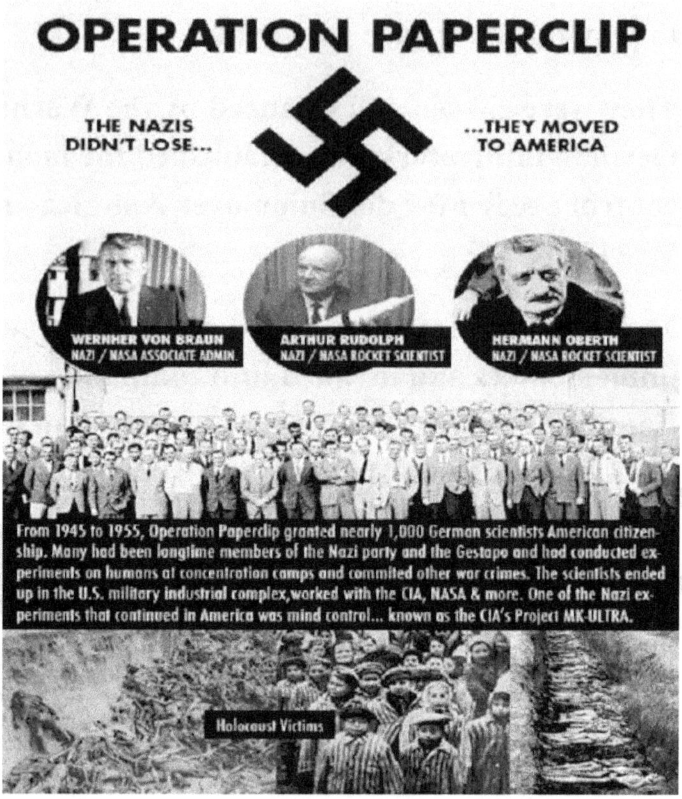
OPERATION PAPERCLIP
THE NAZIS DIDN'T LOSE...
...THEY MOVED TO AMERICA
WERNHER VON BRAUN
NAZI / NASA ASSOCIATE ADMIN.
ARTHUR RUDOLPH
NAZI / NASA ROCKET SCIENTIST
HERMANN OBERTH
NAZI / NASA ROCKET SCIENTIST
From 1945 to 1955, Operation Paperclip granted nearly 1,000 German scientists American citizenship. Many had been longtime members of the Nazi party and the Gestapo and had conducted experiments on humans at concentration camps and commited other war crimes. The scientists ended up in the U.S. military industrial complex,worked with the CIA, NASA & more. One of the Nazi experiments that continued in America was mind control... known as the CIA's Project MK-ULTRA.
Holocaust Victims

It is a great plan, say to America in 1945 that we won the war. The winner take all deal was in action but both sides were funded by the same profiteers and New York Bankers including Prescott Bush.

You do not tell the public that America almost lost if not for weather modification over battlefields, yeah another story.

Then military/industrial companies and NAZI operatives already in all agencies and military bring the planners of the world dominion scheme to the victor nations or sent them to other countries as refugees and there they work quietly to build a Third Reign world system controlled by Germany.

These old money, air money companies also conspired in 1935, to overthrow President Roosevelt and install government by business after the NAZI business model. Retired Marine Smedley Butler was asked to lead the coup charge but loyal to America he testified before congress.

No prosecution was made because, per a source, "it would have devastated the high society façade in New York".

The story of the coup was reported by controlled "fake" media and it was sold as a "gigantic hoax" to the public and they did and still do this today.

Just like 911 an inside false flag job that is never reported (for long) with the evidence that concludes this false flag reality. Like the killing of Kennedy; Like chemical trails delivering sickening agents to you in the air that the space programs of the world are poisoning the air with aluminum fuel exhaust of

space craft is never told with the evidence in hand and in hair without fear.

Same game, same people next generation.

Same people, same agenda that threaten America not the USA façade business deal.

Hitler lost to drugs, fake news and weather modification crippling battle field conditions. His ideas of world dominion are intact and moving forward every second.

But, his New World Bankers, the Warburg bankers won the war cash flow and financed the idea of bringing the enemy's ideology and alien-like technology to America. They, through the power of money and secret religion programed operatives, they put them into tracks of authority that would roll for generations like fifth columnists in the French Revolution.

When Bill Clinton came into office he ask "where did this technology come from", we got no public answer.

With great forethought, they came in lab coats bearing teachers credential to America ready to

spring on democracy and install a new operating system.

A system controlled by corporations and without laws slowing them down. I call it 'madism'. If inbreed acceptable business madness and fascism had a child it would be called the *"New Century of Old World in Dis-order"*

Eisenhower and Kennedy warned us of this situation, just read their speeches. I understand that presidents have no control over the military mind, which is tiny.

Then Senator Johnson told us in his "masters of the universe" speech in congress. He said they were seeking total control of Earth by a few.

> **"From space the masters of the universe would have the power to control earth's weather, to cause drought and flood to change the tides and raise the levels of the sea, to divert the gulf stream and change temperate climates to frigid"**

> **"we now have on record the leaders in the field of science, respected men of unquestioned**

competence, who evaluation of what control of outer space means **renders irrelevant the booking** concerns of

physical officers". What does that mean?

The weather attacks they are doing now are controlled by hundreds in "shorts" (needed comfort for their base in Florida) not millions in uniform

And these presidents were and all others seem helpless to resist the German banking financed invasion of scientists and teachers and many others. They are here, now, at all levels in an autocracy "with a domineering rule or control: a boss who shifts between autocracy, persuasion, and consultation."

They occupy all positions in government. At a low level, some can slow the flow of purchase orders of ink needed to sign checks. Get the drift.

The first critical step, in the overthrow of conscience and reason with specific, well thought out, and tested in prison camps, mind controlling

chemicals/drugs was the changing of the Public Health Service and the fluoridation of the America public.

Edward Bernay, a disciple of his uncle Sigmund Freud, is known as the father of propaganda and mass mind control advertisement. He wrote the book. His greatest achievement was fluoride mind control for sales. He is called the father of impulse buying manipulation in America.

It is written that more money was spent by his German backers selling American authorities on the idea to add the ingredient fluoride into the water supply along with aluminum flocculent and many other chemicals that you drink.

It worked for America is now the devil incarnate and overall dumbed down. Killing in the name of democracy and profit for a few is considered increasingly more normal.

Gangster like stuff, a world mob with a military protection racket and America, English and even Russian air forces willing to spray stuff on the World population. I video this spraying in 8 countries over the last 19 years.

Bernay, was in an early, post WW1, wave of infiltrators with the mission of takeover of the mind of Americas for use in war, buying and mental slavery.

He wrote in his book 'Propaganda', *"The conscience and intelligent manipulation of the organized habits and opinions of the masses is an important element in democratic society. Those who manipulate this unseen mechanism of society constitute an invisible government which is the true ruling power of our country.*

We are governed, our minds are molded, our tastes formed, and our ideas suggested, largely by men we have never heard of. It is they who pull the wires that control the public mind"

He was given an unlimited budget to sale fluoride to stupid officials by these "men we have never heard of" to hedge their marketing bet.

Imagine a metal in liquid form that vibrates at a certain rate when energized by a phone signal, music notes or from one of millions of radio frequencies we are hit with every day and these impulse us to do things. Like buy more on crazy Black Friday.

What if the internal impulse vibration corresponds to one that happen on 'Black Friday" during buying frenzies?

BUY, BUY, BUY

FLOURIDE

ROCKS

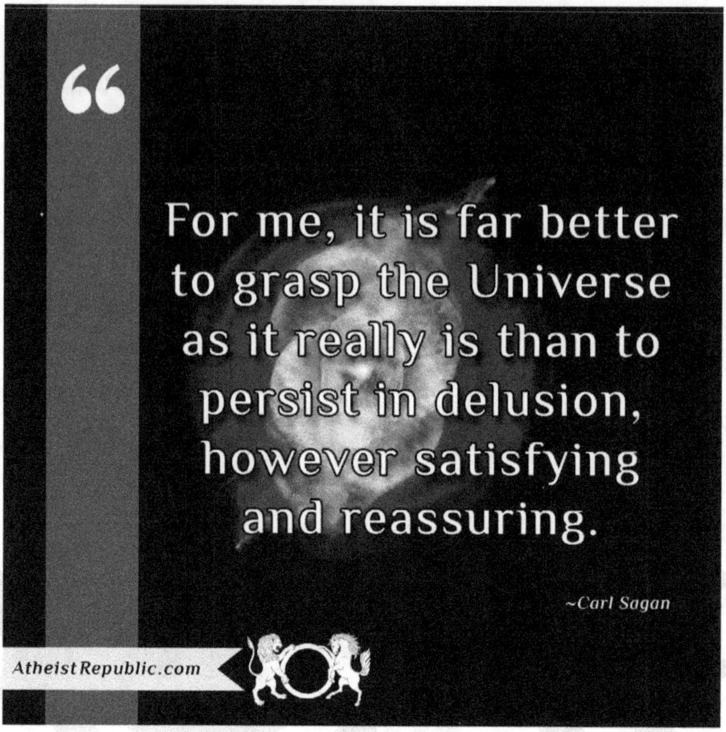
“
For me, it is far better to grasp the Universe as it really is than to persist in delusion, however satisfying and reassuring.
~Carl Sagan
AtheistRepublic.com

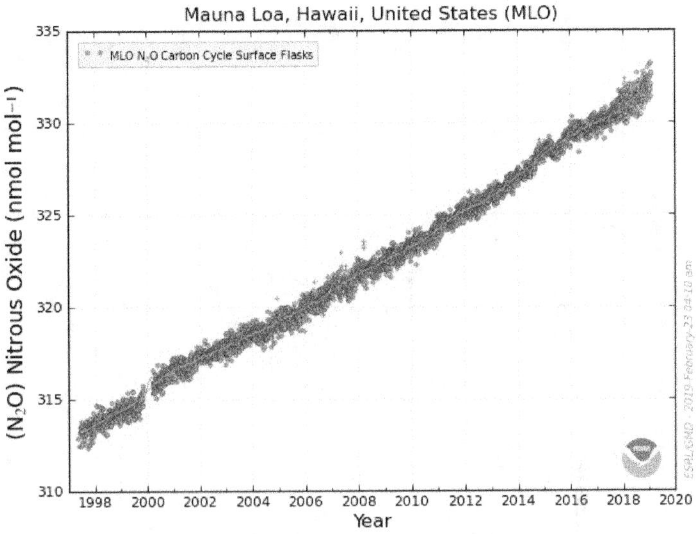

Not eating cattle, or live stock is the big one with, maybe, a planet saving/mitigating affect quickly: Break the carnivorous habit. Buffet please stop singing the cheeseburger song!

We must quit meat or devise grow areas for livestock inside where all gases are captured. Or we soon live like in the movie 'Soylent Green" and pay more than gold for gristle and bone.

DAMN 1950ish Again

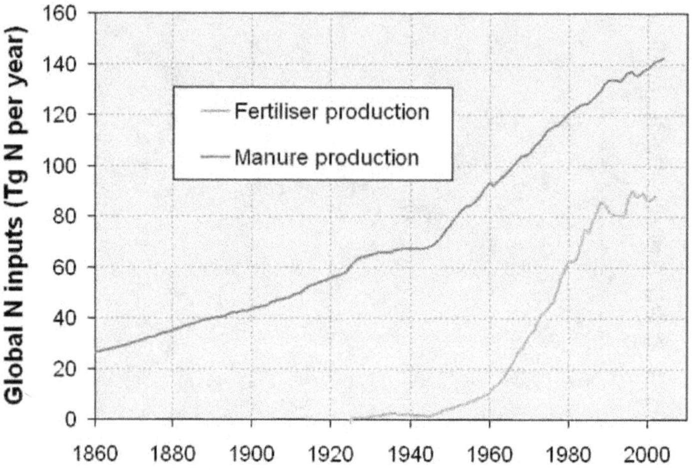

Before 1950 our planet was in a state of slow decline in equalibrum.

The rapid rise in co2 began in 1921. The carbon deposits appeared in skeletons that year.What set off the gluttony of America and its lethal pollution. Drugs, alcohol, a planned mix of gases, and the relief from a national money manipulation "depression". Sound a little engineered yet?

More than half of Americans, any color, are fat. 60% are women . Cattle means mountains of manure and cattle eaten make 'humanure' that steals oxygen and never gives it back.

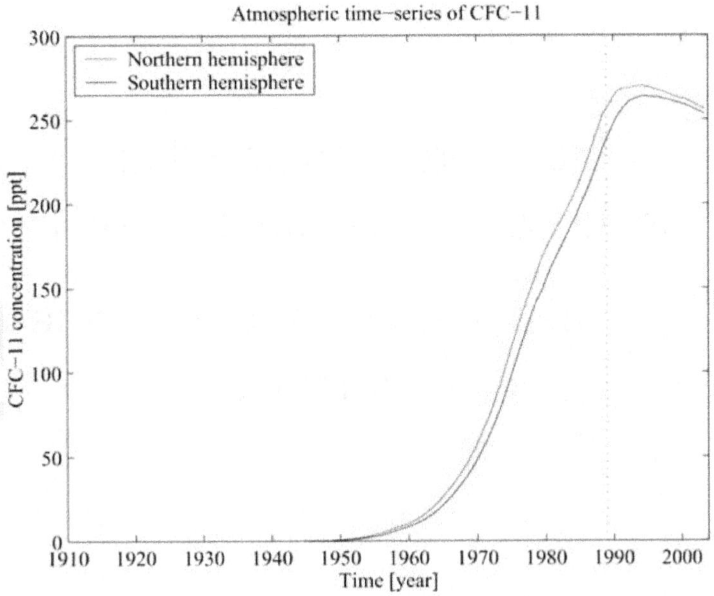

As people got fatter they become less tolerant to heat so the ripple effect builds into a planet killing wave or tusanmni. Thousands are dying in Europe when the heat kicks in. Ok with this so far?

1950

SORRY I MUST HAVE BROUGHT SOME BAD MOJO WITH ME WHEN BORN

Do corporations and their mass advertizing drive meat/blood habits

Chemically or is it programed impulse

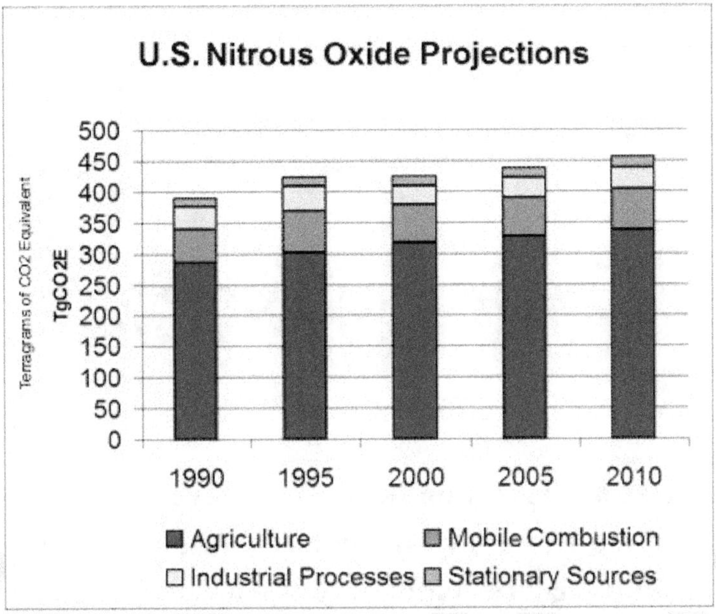

THIS IS LAUGHING GAS AND THIS IS NOT FUNNY

Nitrous oxide is used as a dental anesthesia. At some time without doubt, the level now in our air supply, will take mammals or us to a place of permanent mental numbness.

Maybe we are already there and the madness of war, and mass murder now common are powered by this gas and co2 both mind-altering gas. Co2 is used as an insecticide.

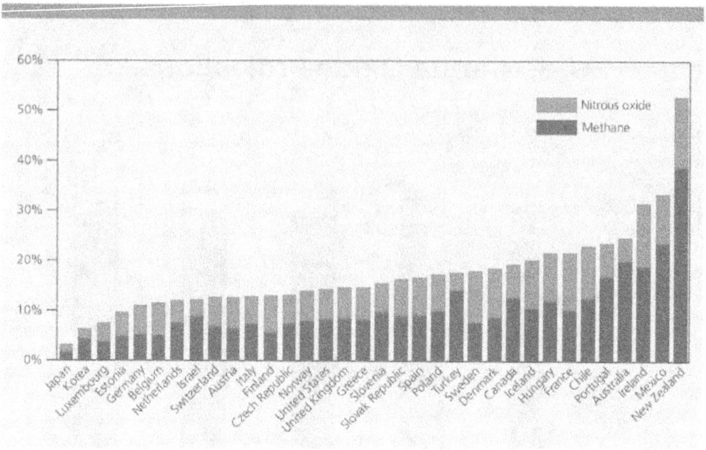

Is there a correlation here: Is the level of laughing gas in the air supply of some regions, or low air flow cities, correspondent to the level of more primitive thoughts and desire to murder for control or fun?

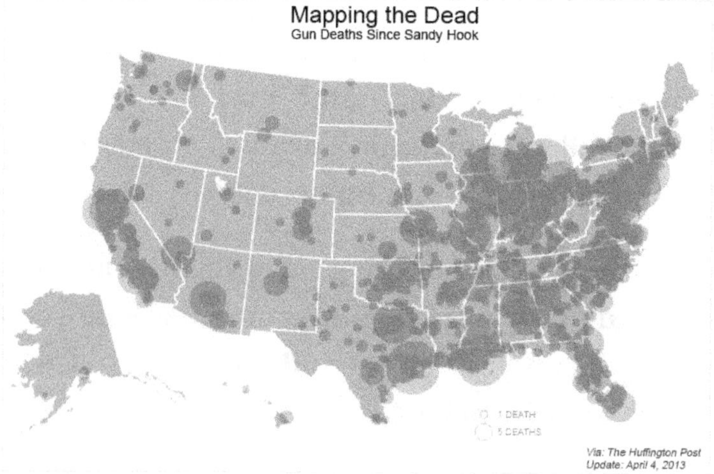

CO2 DRIVEN KILLING: "FOR EACH DEGREE OF TEMPERATURE RISE IN THE USA IS ANOTHER 24,000 MURDERS AND ASSAULTS" WILL OCCURRED, ARE OCCURING".

This study conclusion came from the American Psychological Association

Figure 2. Density Map of 1997 NITROGEN OXIDE Emissions by County

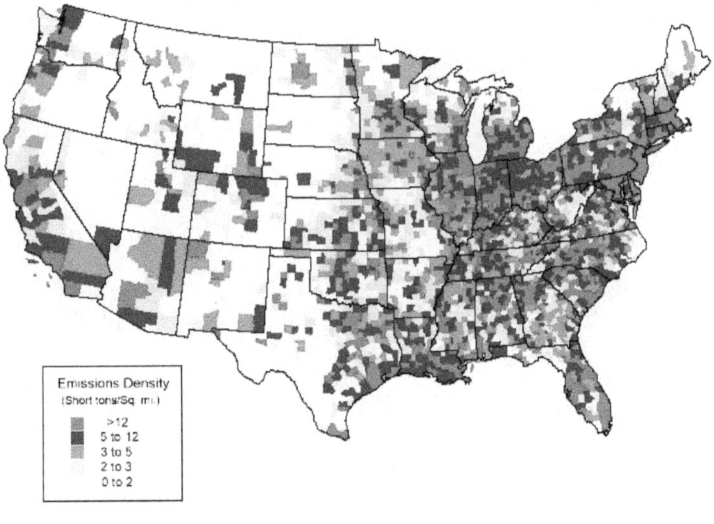

CORRESPONDENCE?

WITH CHEMICALS IN AIR AND WATER

MAYBE DRIVING THE KILLING

Is killing another human ever a natural act?Even in uniform.

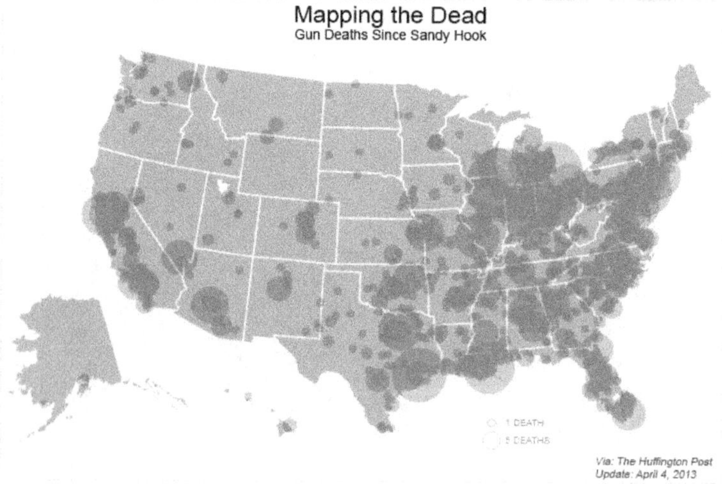

THIS IS SIGNIFICANT: I SHOW IT AGAIN

KILLING IS NOT NATURAL (UNLESS YOU ARE HUNGERY). 'DE-LIFING' IN ANY FORM IS UN-NATURAL AND PRIMATIVE. FROM GROG FOR THE SAILORS TO METH FOR THE SOLDIERS OUR TROOPS HAVE BEEN GIVEN DRUGS TO INSPIRE KILLING AND THEY BRING IT HOME. HERE IS MORE EVIDENCE

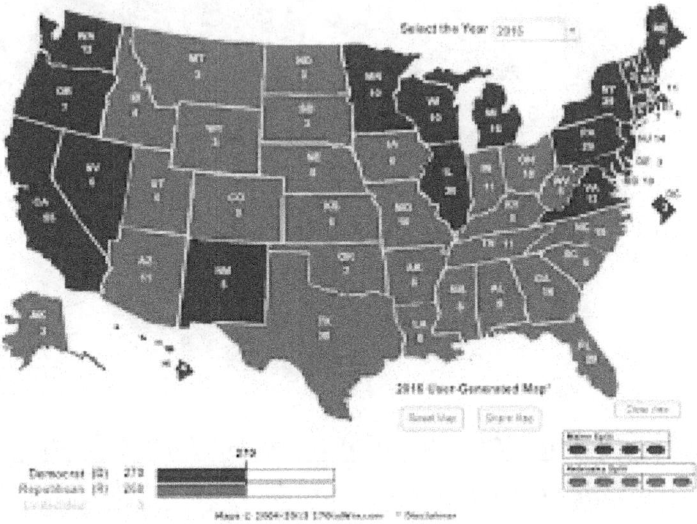

Why are the* RED *states killing states? Maybe more air toxins allowed! Fracking gases can now be measured in the air supply and* BLOOD FLOW *of many of these states. RED STATES SIGNIFY WITH BLACK DRESS.

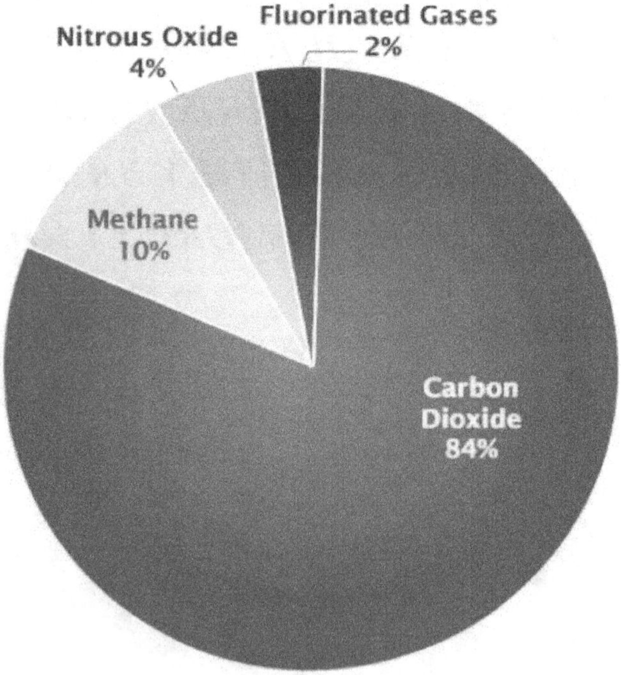

<u>Fluorinated Gases and nitrous: This 6% man-made mind altering gases may be responsible for many</u>

mental dysfunctions around the world.

Co2 is a primary cause of all cell confusion and is one root cause of mammal disease. Displaced essential minerals the second that I understand

Co2 can affect our body electrical system profoundly

We have no other choice and the obstacle to sentient animal cannibalism is lack of understanding that all creatures have feelings and eating them is cannibalism. But not the chicken we had last night, sort of.

We are wired together into one electrical/meta-electrical central nervous/conscience system "hard drive" the magnetosphere.

Here is where memory of all planetary experience resides, as electrons "as above, so below". Well this is interesting, those who invent devices we carry around that emit harmful electrons are the same who understand the correspondence of electrons and human evolution. Getting close.

This manipulation can alter its function and ability as a live planet to remember to evolve. Might forget to change polarities from time to time and change the mind of the planet. Yes, Princeton scientist reveal that reversed magnetic field lava tracks show the planet changes it polarity every 50,000 years and alter all future evolution.

HIGH BRAIN TIDES

Electric shocks running

Electric shocks running

through my brain

Running, running for the life train

Got to go

Got to live

Got to live

Farming at the Meaning

Fields of life reason

Another peak

Needing peaking

Got no more to give

Cannot, not give

In these times

of awareness

Tides rising

Not receding this life time

Last acts not forsaking

Electric shocks

running through my brain

Never, never life is not sane

It takes two

'To get on through

to the other side'

Move on up to the other sides

Magneto spheres

Circle our Earth

Brilliant fools

Hot wiring Heaven

A switch board in the sky

All of us

All of us

Connected

to the hard drive

in the minds of all

Connected

Connected

Electric shocks running

through my brain

Running, running for the life train

Got to go

Got to live

Got no more to give

But if I give

More comes to fill

Connected

Connected

We are one mind

We are all for one and one

for all the many

Just some seed thoughts.

The most important thought I learned from Gurdjieff: "In a relationship each is the custodian of the others soul"

If we, the collective conscience, forget how to evolve and do not remember what is next or past, like a puppy forgetting the tit and dies of starvation with a full breast lined before them and die off.

These controllers share a dream of being the "Masters of the Universe" NASA/NAZA was the vehicle and disease causing aluminum their medical 'profit' benefit program.

We will be replaced by the DNA screwing kings and queens' victors or we live and finish the assignment of evolution to be the best humanity possible.

Changing this energy field would change things in one generation now under attack by hermetic retrogrades, who want to be "Masters of the Universe" who understood the metaphysics.

Need more evidence: This is out of government documents all this madness came from Paperclip Invaders

1950's: Massive Air force

weather modification programs under peaceful military equipment use (but the aluminum and

barium were the human toxins they used to seed clouds). Many lawsuits were filed by farmers and others over major rain disruptions over crop belt. Was this an effort to control food or disrupt the nation with famine.

1952: W.O. Schumann identifies 7.83 hertz resonant frequency of the earth. 1958: Van Allen radiation belts discovered (zones of charged particles trapped in earth's magnetic field) 2,000+ miles up. Violently disrupted in the same year.

This area of energy concentration is also vital to

all mammals. The energy field that is connected to

all life is now changed without thought of the consequence to all living energy creatures.

1958: Project Argus, U.S. Navy explodes 3 nuclear bombs in Van Allen belt. This constitutes an attack on our collective conscience. My thought on this activity is simple: by attacking our spirit circuit man could be under attack by a force seeking to break our will to resist and prevent the universe from answering our requests for assistance with life.

1958: White House advisor on weather modification says Defense Dept. studying ways to manipulate charges "of earth and sky, and so affect the weather

1960: Series of weather disasters begin. 1961: Copper needles dumped into ionosphere as "telecommunications shield".

1961: Scientists propose artificial ion cloud experiments. In 1960's the dumping of chemicals (barium powder etc.) from satellites/rockets began. The silent war began on America, England and other players in the space race business deal controlled by Hitler's elite science troops.

They were now driving the chemical and DNA destiny of all countries who are poisoning with fluoride and where jet exhaust exhales toxins with known disease potential to fill hospitals. Reactive metals cause human cell change!

It will come down to us or them: Period, no question. I wrote a short story 30 years ago called "Cat Wars of 2020". The opening scene was two cattle ranchers come over a hill and see their cattle dead as far as they could see.

Thousands of mothers concerned for the health of the children went on a rampage and killed all the cattle they could find near Amarillo Texas.

I think we are close to mind controlling drug/chemical saturation in the cities and large enclosed areas like airports and government buildings. This saturation must have a significant negative outcome. Mediations now or else.

Move back to country maybe? The same planners of this madness engineered the stealing of 25 million and counting family, securer food sources, small regional farms. For security America, must have a continual web of free harvest farms from Maine to California to Alaska

IN THE PHOTO VON BRAUN A NAZI AND HITLERS REPLACEMENT, IN THE CENTER, CONSPIRED WITH AMERICAN/GERMAN BANKING ELITES TO KILL KENNEDY SIX DAYS LATER.

HE WAS THE FRONT MAN AND THE CONSPIRACY SPEECH WAS ABOUT HIM AND HIS LITTLE FRIENDS. HE AND HIS SLIMY FRIENDS INFILTRATED "INTO EVERY AGENCY OF GOVERNMENT, STATE AND FEDERAL" EISENHOWER

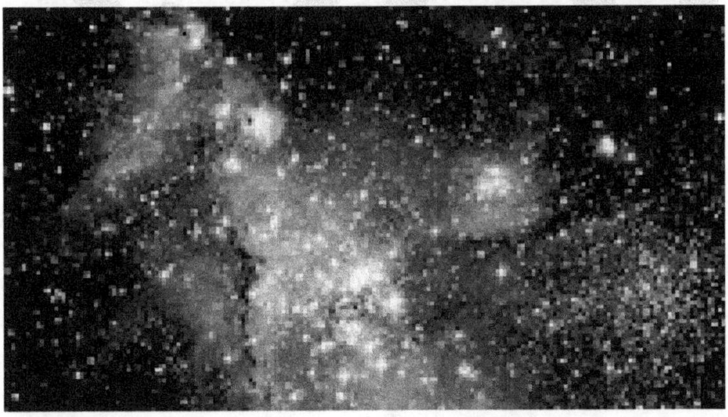

My gathered and very vetted information and video reports from around the world CONFIRM a major weather attack is upon and over America.

Russia and China are seeding the Jetstream before it crosses Alaska. The deep state chemo air force is a constant in the skies of America now in 2019.

That's the cool part as the stuff goes around the Earth and we add our fair share of sky shit, it hits them in the back of their heads and falls into their water. Go figure, but, they may be in on it and this is the plan of the now multi-colored want to be

Masters of the Universe. Maybe they already are our masters? Not: We smart when sober.

Small Farmers are fading away as mega farms become the norm and the sales of toxic chemical that kill, especially farmers, all on the way to the sea, with known disease potential to fill hospital.

That number will be significantly higher when the deception smoke clears after this escalation of control activities and obvious price sabotage actions yield the dividends sought.

They rig the game well: All future fines after profits for its use, an example Round-up, were built into the price.

The chemical Industry, having rigged the game, know some would get fined and

most would not. The chemical industry then prospered by the average. They are all the same golf gamers club so get the drift.

I was in the middle of the fading times of my ocean as a place to make a powerfully rewarding living and feed millions of you.

Price fixing in collusion with Vietnam, China and Equator, and others allowed pond raised and toxic by design shrimp swam into willing corporate chain kitchens. The era of toxic shrimp as a weapon could walk into our bodies, even with protest and evidence enough to stop but know. The noose is very tight these days of consolidation. My film 'Louisiana Falling' nailed that story in 2003.

Alarms now: **In the last year, Diebold, a German origin voting machine company, after getting caught writing rigging programing in half of Americas machines to reverse votes. They sold our voting machine programing to, guess who, a German company with NAZI roots. Guess who they bank with. The smart Germans/bankers, in a move**

to lose the trail, sold the voting machine programing division to a Swedish company with shady roots. Lot of German WW2 sperm invasion adults in business now across Sweden and other blue eyed blond environments.

Alarm: **Monsanto finally losing in court sold all its seed patents for your food crops to, guess who, a German company who sold them quickly in a "strategic move" to China while Trump sat on the toilet and tweeted smoke and mirrors unaware? Not sure but someone advising him should have known.**

I understand that the mind/thought process control objective is the name of todays' slavery. Recent PBS report stated 187 billion per year, 911 overt action, is now spent on keeping track of the flow of life, information, technology used against us the bill payer

The desire to enslave is the primary operating principle, of a few controllers of the money and others of some races, of today on Earth and for thousands of years of plotting to rule over more

clean, less inbred human tribes.

Ask American Indians!

My roots are UK common DNA but I am concerned I may be one of them. To think out this book I must think like them. Just like the idea that everything has two sides or two polarities that allow the paradox of good and evil being the same. I am the opposite.

I am not un-like them. I just think better for the best reasons for the common good.

CHEMICAL WEAPON? Yes! It is displacement of essential MINERALS and elements by specific displacement and reactive chemicals. When mixed in the body they vibrate ***(everything vibrates like music, cannabis concentrates good vibratory energy. Cannabis is here to overcome negative vibes)*** **at a level capable of displacing lower vibratory elements and minerals causing negative system changes in all of us. An answer to many diseases is in the water and it is intentional by my dialectic moving logic.**

And some impulse killing in America Cities may be due to bad air flow in building!

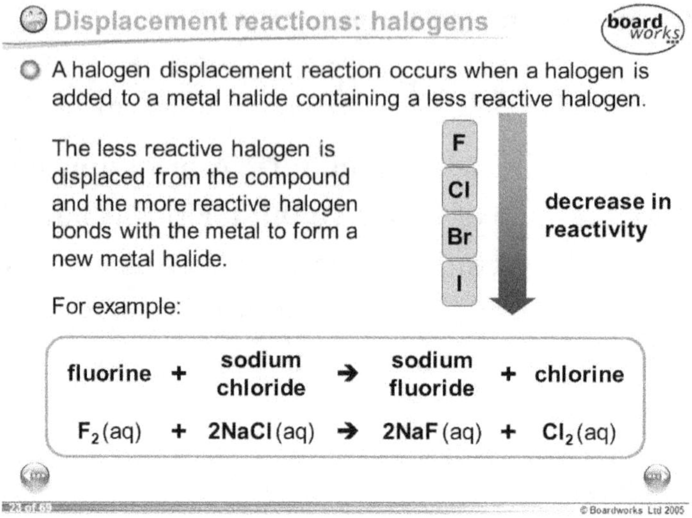
Displacement reactions: halogens
board works
A halogen displacement reaction occurs when a halogen is added to a metal halide containing a less reactive halogen.
The less reactive halogen is displaced from the compound and the more reactive halogen bonds with the metal to form a new metal halide.
F
Cl
Br
I
decrease in reactivity
For example:
fluorine + sodium chloride → sodium fluoride + chlorine
$F_2(aq) + 2NaCl(aq) \rightarrow 2NaF(aq) + Cl_2(aq)$
23 of 69
© Boardworks Ltd 2005

Look up on goggle and read the EPA list of a myriad, 68,000 and counting, of chemicals that end up in our bodies. The unneeded chemicals, for the most part, are part of economic flow and have little to do with health or common sense.

They are in everything you eat, drink, breath and give your babies in your breast milk and continue with formulas designed for future disease. Lack of breast feeding predisposes many doctor visits. Spare the tit get a life of regrets trying to make up for that immunization lack.

Are we numbers and pre-disposed customers for this food/chemical/medical systems' profits.

The transmutations in our bodies of these chemicals, into compounds that cause disease and *customers*, are well thought out by bad scientist likes Edward Teller, Einstein and many Paperclip infiltrators and others.

These "scientific-techno elite" should have been on the list of German public enemies of the planet.

They were German enemy agents brought from behind the lines of enemy countries and they had allegiance to their ethnic and blood birth countries.

We are taught to believe these imported enemy agents/scientists were angles of mercy and not items of hate wanting revenge and control over all other DNA groups.

Their still hot A-bomb dust is still scrambling DNA with radioactive and toxic wind carrying dust that kill thyroids and cause profitable disease. Teller admitted that his radioactive bomb debris would be a good way to weed population of "useless eater" like Kissinger said many times. He should have looked in the mirror.

To know the nature of the enemy and the way they are fighting requires knowing who the enemy may be.

All presidents, are and have been, helpless to this threat of the air money exhaling Germanic creeps.

President Eisenhower in his final speech as president said that the "scientific/techno elite" were a threat to the planet.

Their threat is far above the threat spiritually of generals who use the bombs they design and make the bombs for mass murder and subliminal threat. The mess the environment is in is proof in the pudding.

All wars are for population reduction, profit and mob activity as protection racketeers.

Mob mentality is the causative factor in the looming destruction process happening across the planet. Mob fascism, or corporatism, is the blinder between reality and dividends.

Back to Eddie (mad bomber) Teller:

He pushed the idea of spraying the planet with reflective aluminum Nano powder. Check the following hair sample results and see his idea in every one of us causing at least 58 diseases including Alzheimer's.

He understood molecular physics. NASA is the devil here using aluminum to sicken and profit.

Each shuttle releases 170 tons of aluminum from an antiquated fuel formula. The smarter Russian changed their fuel to reduce aluminum by 2000 times the USA. This seems like the use of a weapon of culling knowing the damage and this is felony.

No thanks to NASA for this over New Orleans during the BP assault.

He, and NASA, knew the unnatural process of ionized metals in our bodies, especially aluminum. Aluminum is highly absorptive and is reactive with other metals and halogens like fluoride, routinely put into your drinking water.

AL is naturally occurring but not in the form in our blood stream. It is processed to kill and exhausted from shuttle rockets and others without regard for public safety

NASA should be shut down and send home all but those support people for those unknown to us human exploration that should, by logic, be happening.

No one can track NASAs' funding of strange science things through a "super crazy science fund".

Teller knew this technique was a great mass subtle disease causative agent and would insure his plan to cull chemically weaken humans he deemed inferior.

Germanic Teller once said in response to a question about atomic bomb fallout speeding up evolution by weeding the weaker ones, he answered, "yes this is a good way to cull Earth of lesser evolved species". He was one of the lesser and thought he the mountain.

America is in the long-term process of ethnic cleansing. Hitler got his model for eugenics from American Indian killers in spirit.

Those in the know, know what antidotes they need to stay healthier.

He knew certain groups would be given the antidote. Look up Wall Street Journal Oct 17, 1997 "Planet Needs a Sunscreen" by Eddie Teller. Mad or marvelous?

This was a cover story, get it, for NAZA/NASA and the reality of their aluminum carelessness. I am so glad most people in government are as hair/body toxic as me. I seek devine retribution by distribution of knowledge and clear thought. Delivered with a smile.

3. Single Displacement (cheater)
One element knocks out ("displaces") another element in a compound

All whales tested have 5000 ppm aluminum in their flesh. We all are taking it in every day and few know to

clean it out to prevent pre-disposed disease for profit. I wonder if this guy who eats same things whales eat is toxic.

YES

In the not too distance past I felt Germanic Teller was a monster but today he may be the man who saved Earth from us. The Manhattan Project bombs destroyed the ozone and a lot more and they got away with it. Maybe his conscience bothered him. Not!

I kind of agree with what I know is happening to cull the planet of what some consider lower defective groups of inbreeds with lots of DNA traps. Hope I am not one but know I am.

Are you one of those chemically dumbed down people being poisoned for subtle, slow and profitable slaughter without much opposition.

I religiously use clean wild spring water that I filter. All other water is toxic to some accumulating degree. I chelate orally often. We are demineralized daily by chemical displacement from air, water and food. I re-mineralize every day and use vitamin d 10-20 thousand. Have too or else.

None of this is an accident. Keep reading!

I am lucky because I know why I am on Earth now doing what I do.

My evolution facilitating job is to assist the process of developing, with information and logic, a more aware people.

Yes, I assist evolution to generate a more evolved next generation. Wish there were more like me. More who heard the storm warnings warning. There are many who tried, but few who stay the route when the route is not clear. Like a drowning person saying "I know I am drowning and I will try to swim. Understand the meaning.

Evolution happens with each generation or that was the plan until this generation when mind control became the game of sub-life for a few mad dogs hiding inside mountains cities.

Tunnels under the USA and Deep Underground Military Bases (DUMBS)

See also, Underground Facilities - Bases - Tunnels

So the elite are keeping silent about Planet X coming but have been busy building underground cities and interconnected tunnels - hmmmmmm? Don't believe that? Keep reading.

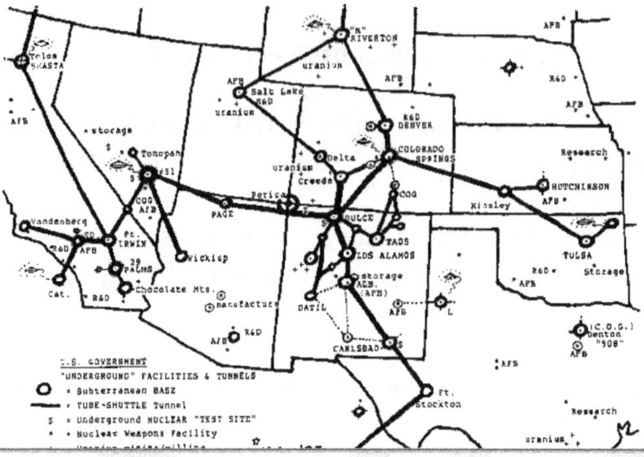

Teller may be an instrument to assist the fittest to survive when ignorance kills off the competition.

I may be an instrument for that change: Are you?

Those who read my books and take the health advise will

Live Longer **with less disease downtime or early**

death. Knowledge is power over ignorance.

Ultra-state operatives know there is long-term population culling underway. (Goggle Agenda 2100 and MK ultra, then do your own research)

Population reduction **is a "natural action"**

an example maybe war. Some wars today are called by world press "ethnic cleansing".

This is never fully explained.

I see it as DNA culling like filtering the chaff from the wheat. Clearing the playing field of defectives is a good plan just the defective seem to be making the decision about who is defective. See the problem we have on Earth.

This "ethnic cleansing" is making the way clearer for a cancer like DNA group to occupy Earth in the future.

This is the nature of planetary war.

Yes, this evil "force" is across the universe. 'Star Wars' got close to the reality of time and Space Wars.

Space war hummm-A double meaning maybe.

Of note: Population reduction has been seen among other animals. When too many cattle are placed in a pen, some will fall voluntarily and be killed to make room even while waiting to be eaten.

Our American, secret to most, reality today: A decision was made after the Global 2000 report came out in July 1980 or later. It was ordered in 1977 and authored by President Carter.

At that time Paul Ehrlich, a Germanic, recommended for a fair and equitable reducing of planet population to the administration.

I did my own 'Global 2000' research. I read the books. Two of them. One was technical and the other a summation of 135 experts in all fields of science and more.

In summation, it said 20 years ago: we are fucked if we do not change our course of planetary management now. We did not change anything except speed of destruction leading to a pre-determined location called hell on earth.

Smell the sulfur yet or lack of it. I do, yes I do.

I did my 31-year search/research to write these books, first, for my own knowledge and information

I needed for future action.

Future is now this book is reality. And no one died telling you this yet. All of us are dying before time is ready if we do not act.

I needed to understand how a matter of fact decision could be made to reduce the planet population of the human herd.

It made sense then and more so now. But I still could never be the one to say kill half the population. Not my job

description or not offered enough money but could happen.

Population culling is in effect now and the ones who made that decision are going underground or on super ships like in that movie.

When information manipulators make a movie like "The Day After Tomorrow" it is already happening and this creates a plausible denial response concerning this reality being a now possibility. Calm the masses. Like Bush said on 911, "go shopping, go to Disney World".

There are 7 'Super Ships' already built complete with zoos and cryogenics egg and sperm storage on each ship. This is real I saw the first Armageddon escape ship 10 years ago when it came to Florida to install the zoo and board the animals. Sound familiar.

My explanation is simple kill cleaner less inbreed DNA strains and over time a dominate DNA group will emerge, like cancer, and win the day and the planet. This is happening across the mega-verse.

Planet Infestation: To be the one to rule all others, to infest and control all food supply and all mammals, is the slaver dream of yesterday and now.

These system controllers know that few chemists, with guts to speak out, will not, because most work for the system is a form of silence servitude for the grants.

The few thoughtful researchers who know the big truths I am showing you rarely speak out. NASA/NOAA front man Hanson was an exception or he was the guy sent to set-up the conversation to prepare population for a bigger fall.

Recently, my spine surgeon, after seeing my films, 'Smoky Truth' and 'Vitamin D Deception" about this situation called me a "genius" because I have out thought thousands of paid researchers. Why aren't I rich?

I understand that single subject researchers are incapable of seeing this planet population reducing toxic pollution puzzle.

STAGES OF ADAPTATION

1 ALARM

2 RESISTANCE

3 EXHAUSTON

*SO, **FUCK IT**, GO ON*

DO IT

VACCINE CONTAINS
ALUMINUM

WAS IT GOMER PYLE WHO SAID, "SURPRISE, SURPRISE"

ALUMINUM DISPLACES IODINE AND CONFLICTS VITAMIN D

GETTING WARM YET!

ALUMINUM AND FLUORIDE IS IN ALL

BREAST MILK

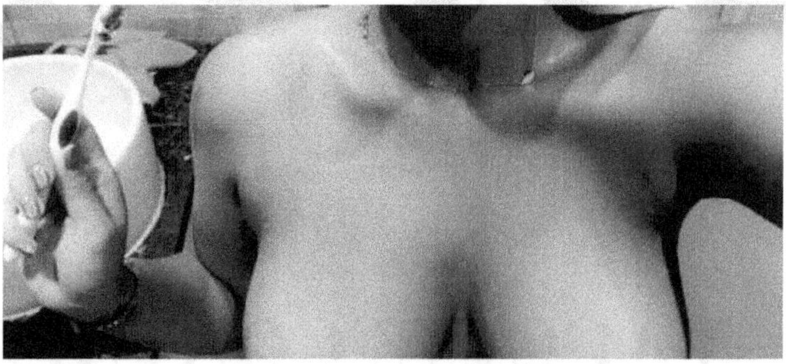

THANKS TO NASA/NAZA

1965 Vaccine Schedule

Tuberculosis, Polio and Tetanus

1983 Vaccine Schedule

24 Doses

7 Injected Vaccines

4 Oral Vaccines

DTP (2 MONTHS)

OPV (2 MONTHS)

DTP (4 MONTHS)

OPV (4 MONTHS)

DTP (6 MONTHS)

MMR (15 MONTHS)

DTP (18 MONTHS)

OPV (18 MONTHS)

DTP (4 YEARS)

OPV (4 YEARS)

TD (15 YEARS)

1986 ***WAS A GOOD***

YEAR PHARMACEUTICAL MANUFACTURES PRODUCING

VACCINES WERE FREED FROM ALL LIABILITY RESULTING FROM VACCINE INJURY OR DEATH BY "CHILDHOOD VACCINE INJURY LAW"

2016

74 Doses

53 Injected Vaccines

3 Oral Vaccines

INFLUENZA (PREGNACY)

TdaP (PREGNACY)

Hep B (BIRTH)

Hep B (2 MONTHS)

Rotavirus (2 MONTHS)

DTaP (2 MONTHS)

HIB (2 MONTHS)

PCV (2 MONTHS)

IPV (2 MONTHS)

Rotavirus (4 MONTHS)

DTaP (4 MONTHS)

HIB (4 MONTHS)

PCV (4 MONTHS)

IPV (4 MONTHS)

Hep B (6 MONTHS)

Rotavirus (6 MONTHS)

DTaP (6 MONTHS)

HIB (6 MONTHS)

PCV (6 MONTHS)

IPV (6 MONTHS)

Influenza (6 months)

Influenza (7 months)

HIB (12 MONTHS)

PCV (12 MONTHS)

MMR (12 MONTHS)

Varicella (12 MONTHS)

Hep A (12 MONTHS)

DTaP (18 MONTHS)

Influenza (18 MONTHS)

Hep A (18 MONTHS)

Influenza (30 MONTHS)

Influenza (30 MONTHS)

DTaP (4 YEARS)

IPV (4 YEARS)

MMR (4 YEARS)

Varicella (4 YEARS)

Influenza (5 YEARS)

Influenza (6 YEARS)

Influenza (7 YEARS)

Influenza (8 YEARS)

Influenza (9 YEARS)

HPV (9 YEARS)

Influenza (10 YEARS)

HPV (10 YEARS)

Influenza (11 YEARS)

HPV (11 YEARS)

TdaP (12 YEARS)

Influenza (12 YEARS)

Meningococcal (12 YEARS)

Influenza (13 YEARS)

Influenza (14 YEARS)

Influenza (15 YEARS)

Influenza (16 YEARS)

Meningococcal (16 YEARS)

Influenza (17 YEARS)

Influenza (18 YEARS)

LEGEND:

DTP: Diphtheria and tetanus toxoids and acellular pertussis vaccine adsorbed (P)

MMR: Measles, mumps and rubella vaccine

OPV: Oral Polio

Hep B: Hepatitis B

DtaP: Diphtheria and tetanus toxoids and acellular pertussis

adsorbed and inactivated poliovirus vaccine

HIB: Haemophilus influenzae type b conjugate vaccine

PCV: Pneumococcal conjugate vaccine

IPV: Inactivated Poliovirus

Varicella: Chicken Pox

Td: Tetanus and dipphtheria toxoids adsorbed (A)

Tdap: Tetanus toxoid, reduced diphtheria toxoid and acellular pertussis vaccine, adsorbed (A)

HPV: Human papillomavirus vaccine

Rotovirus: RV

Influenza: IIV

Varicella: VAR

Meningococcal: MenACWY-CRM

HEAR ME YET?

MAYBE THE CHEMICAL INVASION I HAVE TOLD YOU ABOUT IS CAUSING OUR COLLECTIVE DISEASE

OR

IS THE RISING DISEASE RATES CAUSED BY THE CHEMICAL CURE?

KEEP READING

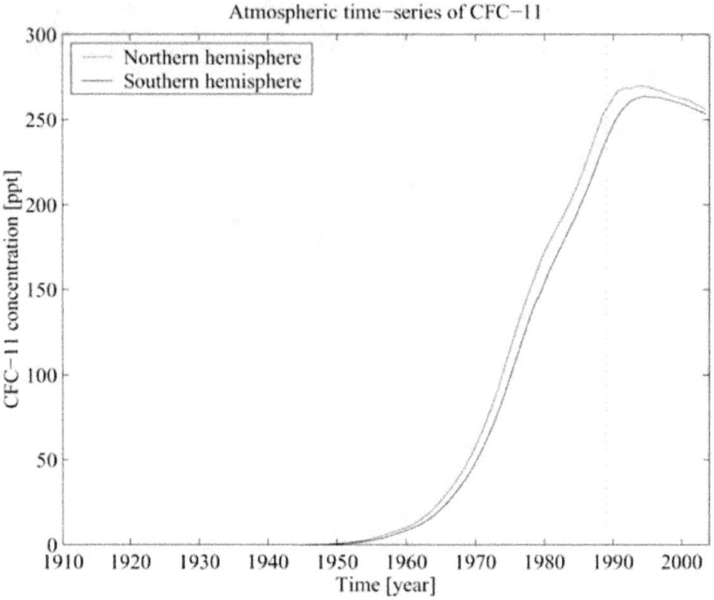

Table E: Solubility of Ionic Solids. (aq) = aqueous, soluble in water; (s) = solid, does not dissolve in water.

Ions	Acetate	Bromide	Carbonate	Chlorate	Chloride	Fluoride	Hydrogen Carbonate	Hydroxide	Iodide	Nitrate	Nitrite	Phosphate	Sulfate	Sulfide	Sulfite
Aluminum	s	aq		aq	aq	s		s	—	aq		s	aq	—	
Ammonium	aq	aq	aq	aq	aq	aq	aq	—	aq	aq	aq	aq	aq	aq	aq
Barium	aq	aq	s	aq	aq	s		aq	aq	aq	aq	s	s	—	s
Calcium	aq	aq	s	aq	aq	s		s	aq	aq	aq	s	s	—	s
Cobalt(II)	aq	aq	s	aq	aq	—		s	aq	aq		s	aq	s	s
Copper(II)	aq	aq	s	aq	aq	aq		s		aq		s	aq	s	
Iron(II)	aq	aq	s		aq	s		s	aq	aq		s	aq	s	s
Iron(III)	—	aq			aq	s		s	aq	aq		s	aq	—	
Lead(II)	aq	s	s	aq	s	s		s	s	aq	aq	s	s	s	s
Lithium	aq	aq	aq	aq	aq	aq	aq	aq	aq	aq	aq	s	aq	aq	aq
Magnesium	aq	aq	s	aq	aq	s		s	aq	aq	aq	s	aq	—	aq
Nickel	aq	aq	s	aq	aq	aq		s	aq	aq		s	aq	s	s
Potassium	aq	aq	aq	aq	aq	aq	aq	aq	aq	aq	aq	aq	aq	aq	aq
Silver	s	s	s	aq	s	aq		—	s	aq	s	s	s	s	s
Sodium	aq	aq	aq	aq	aq	aq	aq	aq	aq	aq	aq	aq	aq	aq	aq
Zinc	aq	aq	s	aq	aq	aq		s	aq	aq		s	aq	s	s

precipitate

Double Displacement Reactions

Barium nitrate + (aq)
Potassium carbonate (aq)

Potassium nitrate (aq)
+
Barium carbonate (s)

Driving force

The Art of Deception

Few understand the art of deception in the use of food additives and the odds they give of how many die, to benefit many in the *known poisoning* for food safety and how they change the

chemistry in our bodies

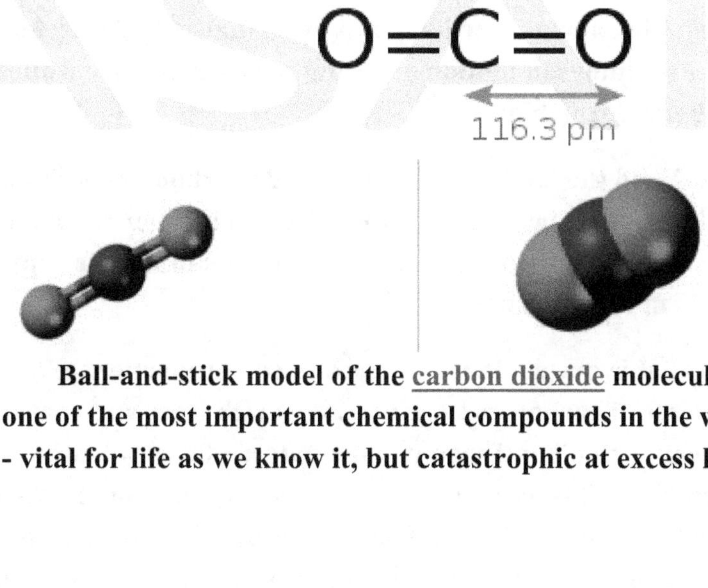

Ball-and-stick model of the carbon dioxide molecule, one of the most important chemical compounds in the world - vital for life as we know it, but catastrophic at excess levels.

An ancient understanding, "As above, so below" may give one the insight to understand how things work on earth and in the 'Megaverse'.

I understand this as saying to us with 'ears to hear' this truth that everything is big and small at the same time somewhere in the system of universal functions.

If there is a big movement in nature there is also a smaller, corresponding entity. Example: A woman's menstrual cycle, if she is healthy, is 28 and one quarter day. The moon cycles are 28 and one quarter days.

Thinking along these lines and using dialect logic (seeing clearly things in motion or seeing the clear picture while others see the blur).

CO2 and the locking of oxygen inside carbon gases will kill all mammals in time from the effects of a hyper carbonic state and steady oxygen supply reduction of the life support system of the planet.

Available oxygen is sequestered in co2 proportionate to release of carbon gases. Cut oxygen rich trees and coat the ocean with oil and soon no adequate air for clear mind function. This happens in closed spaces where many people are breathing the air and co2 builds up in the room. Co2 poisoning is often seen as being drowsy in a room with an open fire.

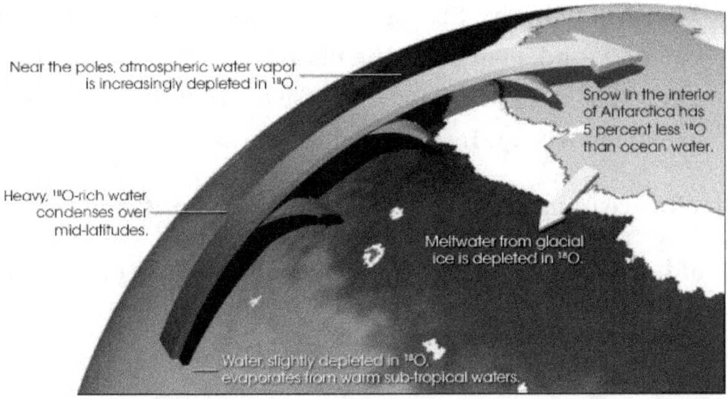
Near the poles, atmospheric water vapor is increasingly depleted in ^{18}O.
Snow in the interior of Antarctica has 5 percent less ^{18}O than ocean water.
Heavy, ^{18}O-rich water condenses over mid-latitudes.
Meltwater from glacial ice is depleted in ^{18}O.
Water, slightly depleted in ^{18}O, evaporates from warm sub-tropical waters.

If we stay the co2 release course our blood chemistry will become acid, just like the ocean has become acid as our bodies try to adjust.

From 1969 till now 2019 the temperature of the ocean, down to 2300 feet has warmed .4 degrees over all the North Atlantic 1.4 Fahrenheit degrees of ice melting radiating heat from the hot tropic via the Gulfstream. A giant meandering river of a current.

Once predictable but now losing direction and strength to the degree of alarm especially England. The British Isles and Ireland will have more catastrophic weather shifts, like happening everywhere now, if the Gulfstream stops and turns right toward Europe and Africa. Look it up!

The mountains of Arkansas and southern Missouri are the main 'safer zones' for those not going underground by plan or lack of invitation.

There has been an exodus by NOAA and NASA scientist and families to these mountains for decades.

They know where the high water and higher oxygen levels will be found soon after the planet shift NOW underway. Yeah this seem real crazy. Please do your own research and prove me wrong. Dare YOU!

The ocean has absorbed so much co2 that it is in the process of becoming acidic. This is the course we are on as small oceans symbiotic to the planet.

We are on the edge of become fatally acidic and many are already there but still functioning.

On the news, just now 6:18 pm May 13, 2019 a report that I saw online yesterday put co2 level in Hawaii has suddenly risen to 414ppm. In one day, it rose to 415 ppm. Is the end near? Yes? What can we do? Stop all pollution to the air by 50% now. Never fly a jet half full etc.

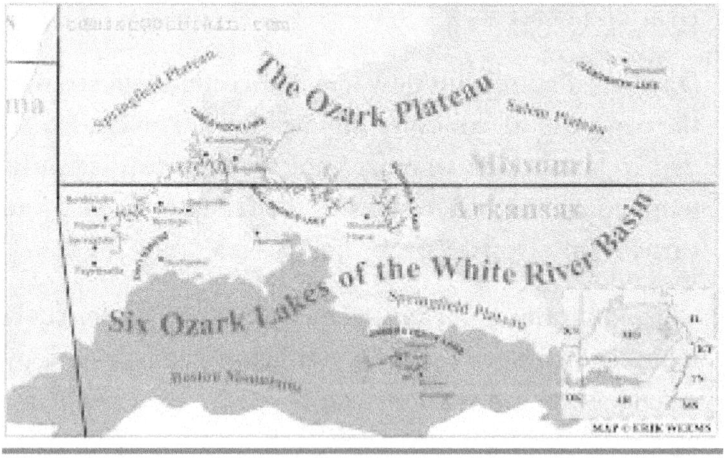

<u>This is a map of exodus of retired NOAA people. They know where the high water will be and the little food. Yeah Arkansas is Noah's destination</u>

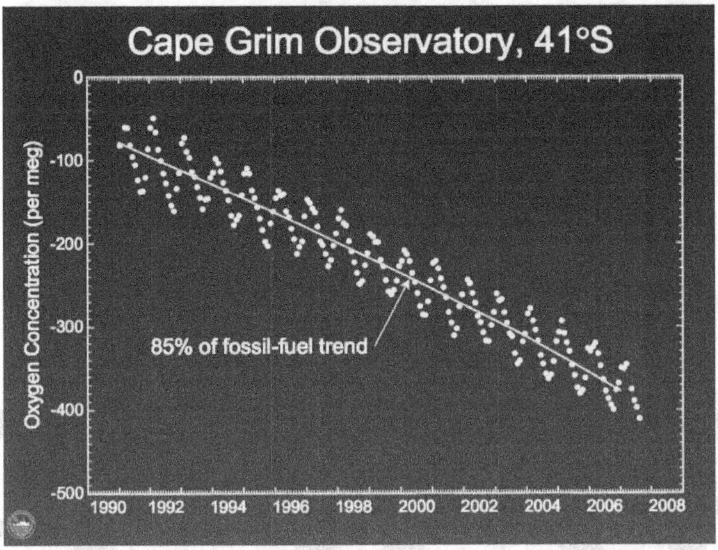

COULD NOT FIND ONE FROM 1950

GET THE PROBLEM CO2 LOCKS OXYGEN molecules 2 TO 1

YOU LOSE IT IS NO LONGER AVAILABLE TO US TO BREATH!!!!

FLOURIDE/ALUMINUM/CO2

LAUGHING GASE? YEAH!

Perfect Storm of Mind Control

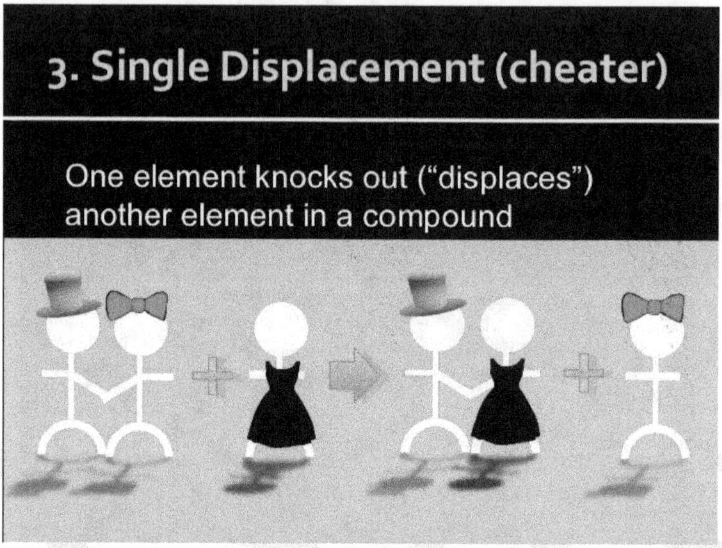
3. Single Displacement (cheater)
One element knocks out ("displaces") another element in a compound

Gas. Carbon dioxide, whose chemical formula is CO2, is a heavy, odorless, colorless, non-combustible, non-toxic (but asphyxiating) gas. A level lower than we are breathing can alter minds chemically and change thought pattern from love to hate maybe.

WOW Carbonic Snow

"It is a component of the air in the atmosphere at sea level in a proportion of 0.02% in 1950, 'now it is at 0.004%'.

34% by volume in very breath and rising daily basis.

Carbon dioxide is soluble in water and alcohol; it becomes liquid at five atmospheres of pressure and at -56 degrees Celsius. Under normal pressure, part of it evaporates and part becomes solidified forming dry ice (solid carbon anhydrous or carbonic snow), used to maintain low temperatures: -89 degrees Celsius.

This gas is used in the preparation of effervescent solutions in the field of Medicine, in fire extinguishers; and, in its solid form, in the processing of clouds, with the intention of modifying the cloud's water droplet size, and to provoke its precipitation or dissipation.

CO_2 constitutes an indispensable nourishment for plants, being absorbed because of cellular respiration.

Retention*: *This gas has been administered, through mask, to psychoneurotic patients with the purpose of giving therapy, creating an ample variety of subjective sensory phenomena, extremely like the phenomena found in near-death experiences.

It also creates the exact sensation of the detachment of the consciousness outside the human body, and self-bilocation.

This proves that the retention of carbon dioxide in the brain, when the individual is exposed to extreme hyper-carbonic conditions, can produce a near-death experience or a forced, conscious out-of-body projection.

THIS EXPLAINS THE DUMBING DOWN OF PEOPLE WITH OXYGEN STARVED MINDS. THIS IS NORMAL IN MOST TRAFFIC JAMS. IS THIS THE CAUSE OF 'ROAD RAGE'?

Alterations: Modifications in the breathing speed and intensity can influence the cardiac rhythm and the arterial pressure, altering the

oxygen content, carbon dioxide, acids, alkali, lactates, and the calcium contained in the circulatory stream.

Thus, it can affect the normal functioning (neurophysiology) of the cerebral hemispheres, either critically or harmlessly.

Symptoms: Hypoxia (the lack of oxygen in the tissues provoked by a low or insufficient level of oxygen), and the hypoxemia (the lack of oxygen in the circulatory stream resulting from, for instance, a decrease in atmospheric pressure) constitute forms of oxygen starvation producing symptoms of asphyxia, suffocation, head buzzing, lack of muscle coordination, visual alterations, vertigo, excessive sweating, emotional unbalance, lack of critical judgment, hallucinations and other xenophrenic states.

Hypo-aeropathy. Each organ in the human body has a certain level of varied tolerance towards hypoxia. The deficiency of oxygen in the inhaled air causes alterations named hypo-baropathy, altitude-sickness, aviator's sickness, etc. This phenomenon occurs in human beings when at great heights, above six thousand meters, in mountains, airplanes, etc.

Warning. The normal air in the room of a practitioner of out-of-body projections contains 21% of oxygen and practically 0% of carbon dioxide. It is known that carbon dioxide (CO2), when inhaled pure (100%), causes the death of the human body due to asphyxia or suffocation.

Phenomena. Six events occur, in the Project-iology field, due to the low level of oxygen in the lungs and because of the increasing concentrations of carbon dioxide in the tissues: rhythmic breathing technique; near-death experiences; accidents due to asphyxia; conscious out-of-body projections in penitentiaries; inadequate sleeping habits; voluntary "mini-deaths".

Respiration. Hypercarbia, or the increase of carbon dioxide in the brain, explains the mechanism by which the rhythmic breathing technique works, as part of yoga breathing exercises. When these exercises are practiced systematically, they cause a prolonged breathing suspension, i.e., the retention of the exhaling, or a diminishing of the breathing rhythm.

This leaves the practitioner with a slight thirst for air or in a condition of voluntary discomfort.

Near-death: Carbon dioxide is normally formed in the brain as a final product of the brain's cellular metabolism. The blood that carries the oxygen to the brain (pure and rich in oxygen) is also responsible for carrying carbon dioxide (impure blood composed of CO_2) from the brain to the lungs, where, at last, it is expelled from the human body. The ceasing of the flow of pure blood provokes a heart attack, or hyper-carbonized brain (hypercarbia), just like a great number of near-death experiences, and the separation of the consciousness from the human body (OBE) in certain opportunities.

Accidents: The increase of carbon dioxide occurs spontaneously with certain frequency in serious accidents causing suffocation or asphyxia. This can also produce a conscious projection in the individual.

Solitary:

The same process of the increase of carbon dioxide produces the involuntary conscious or unconscious out-of-body projection in individuals (or inmates) confined in totally restrictive institutions, in prisons' isolation cells, or in restricted cubicles, where the air is polluted and rarefied of oxygen.

Covers:

The unhealthy habit of people who sleep covering their heads with sheets, which diminishes their capacity of oxygen intake and increases the volume of carbon dioxide in the air surrounding their face and nostrils, can produce, in certain cases, the lucid out-of-body experience.

The use of carbon dioxide shows a similarity between the phenomenon of the conscious out-of-body projections and the experience of biological death. When carbon dioxide is inhaled in its pure form, it produces the death of the human body; inhaled in a small per volume amount (30%), predisposes the projection of the consciousness through the psychosoma (OBE)."

For you techies and non-believers

Carbon Dioxide and Carbonic Acid

THE FOLLOWING IS SOME OF THE SCIENCE OF HOW CO2 INFILTRATES WATER BOTH SALT AND FRESH.

MY* epiphany*: THIS IS THE PROCESS WHERE CO2 ENTERS OUR BLOOD AND BECOMES CARBONIC ACID AND THROUGH A PROCESS BECOME CALCIUM CARBONATE AND BECOMES THE BUILDING MATERIAL OF SPINAL AND JOINT MISERY.

CO2 maybe the glue that allows deposits that create Alzheimer's

FOR THE GEEKS

Yes, redundancies for dummies

"It a fine line" Warren Zevon

The most common source of acidity in water is dissolved carbon dioxide.

Carbon dioxide enters the water through equilibrium with the atmosphere

CO_2 (aq) « CO_2 (g)

and biological degradation/photosynthesis involving organic carbon, {CH_2O}

{CH_2O} + O_2(aq) « CO_2 (aq) + H_2O

Aqueous CO_2 (aq) also undergoes a number of important inorganic equilibrium reactions. First, it can dissolve limestone

$CaCO_3$ + CO_2 (aq) + H_2O « Ca^{2+}(aq) + 2 HCO_3^- (aq)

Second, it can react with the water to form carbonic acid

CO_2 (aq) + H_2O « H_2CO_3 (aq)

Only a small fraction exists as the acid

$$K = \frac{[H_2CO_3(aq)]}{[CO_2(aq)]} = 1.3\times10^{-3}$$

and the kinetics to form H_2CO_3 are relatively slow (on the time scale of seconds).

Carbon Dioxide and Carbonic Acid-Base Equilibria

Dissolved CO_2 in the form of H_2CO_3 may lose up to two protons through the acid equilibria

H_2CO_3 (aq) « H^+ (aq) + HCO_3^- (aq)

HCO_3^- (aq) « H^+ (aq) + CO_3^{2-} (aq)

The equilibrium equations for these are labeled as "1" and "2" hence

$$K_{A1} = \frac{[H^+][HCO_3^-]}{[H_2CO_3]} = 2.00\times10^{-4}$$

$$K_{A2} = \frac{[H^+][CO_3^{2-}]}{[HCO_3^-]} = 4.69\times10^{-11}$$

To account for the fact that CO_2 (aq) is in equilibrium with H_2CO_3 (aq), the first acid equilibrium is normally given by

$$K_{A1} = \frac{[H^+][HCO_3^-]}{[CO_2(aq)]} = 4.45\times10^{-7}$$

The acid equilibrium equations can be solved to give the fraction of carbonates in a particular form.

$$\alpha_{H_2CO_3} = \frac{[H^+]^2}{[H^+]^2 + [H^+]K_{a1} + K_{a1}K_{a2}} = \frac{[H_2CO_3]}{total\,CO_2(aq)}$$

$$\alpha_{HCO_3^-} = \frac{[H^+]K_{a1}}{[H^+]^2 + [H^+]K_{a1} + K_{a1}K_{a2}} = \frac{[HCO_3^-]}{total\,CO_2(aq)}$$

$$\alpha_{CO_3^{2-}} = \frac{K_{a1}K_{a2}}{[H^+]^2 + [H^+]K_{a1} + K_{a1}K_{a2}} = \frac{[CO_3^{2-}]}{total\,CO_2(aq)}$$

Relative H_2CO_3 concentration is CO2 (aq) in equilibrium with water.

In summary;

- **CO_2 enters water through interface with the atmosphere and the biological processes of organic carbon digestion and photosynthesis.**

- **Aqueous carbon dioxide, CO_2 (aq), reacts with water forming carbonic acid, H_2CO_3 (aq).**

- **Carbonic acid may lose protons to form bicarbonate, HCO_3^-, and carbonate, CO_3^{2-}. In this case the proton is liberated to the water, decreasing ph.**

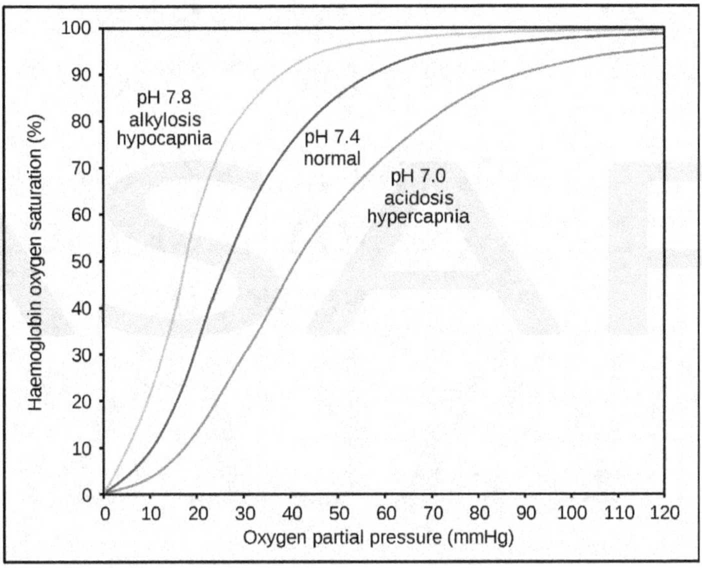

The complex chemical equilibria are described using two acid equilibrium equations.

The first acid equilibrium constant accounts for the CO_2 (aq) - H_2CO_3 (aq)

equilibrium. It consequently seems to have a high pu$_{kka}$.

The fraction of the inorganic carbon in a form is call the "alpha" and there is simple equation to describe this alpha.

THE POWER OF EVIDENCE DIALECTIC LOGIC

IN 1921 THE PLANET OR THE OCEAN WAS NO LONGER ABLE TO ABSORB THE

EXCESS

CO2 RELEASED BY MANS' ACTIVITIES BEYOND NORMAL AND ANTI-SYMBOTIC

THIS GRAPH IS A 'HOCKY-STICK'

Global Trends in Major Greenhouse Gases to 1/2003

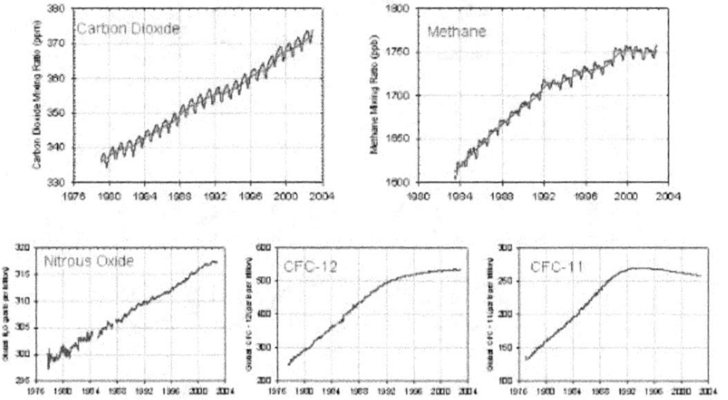

Global trends in major long-lived greenhouse gases through the year 2002. These five gases account for about 97% of the direct climate forcing by long-lived greenhouse gas increases since 1750. The remaining 3% is contributed by an assortment of 10 minor halogen gases, mainly HCFC-22, CFC-113 and CCl_4.

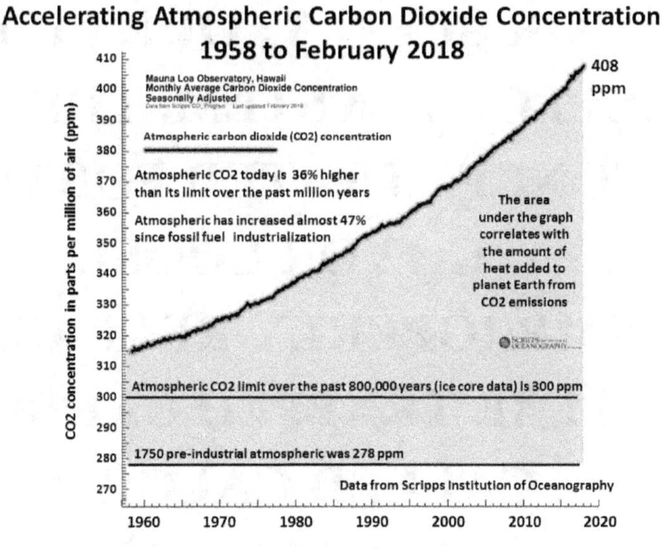

1958

THE SUDDEN RISE IN CO_2 CONTINUES.

THERE IS A BI-PRODUCT OF THE RISE OF C02 IN OUR AIR SUPPLY; OUR LIFE-SUPPORT SYSTEM. THIS IS A CONCENTRATED FORM OF CO_2 OR LOOSE IMPROPERLY RELEASED TIME TRAPPED CARBON CALLED CALCIUM CARBONATE.

THIS CARBON BONE MATERIAL THAT BEGAN AS THE CARBON BUILDING MATERIAL, AND MNERALS, THAT FORMED OUR BODIES AS WE EVOLVED/CREATED INTO HUMAN FORM. THIS EXCESS CARBON IS OVER FLOODING ALL MAMMALS BODIES AND IS DEPOSITING IN OUR SPINES AND JOINTS.

I MAY HAVE CHANGED THE GAME ON PLANET EARTH WITH THIS INFORMATION AND POSSIBLE INSIGHT THAT I HAVE PRESENTED

ON EARTH FOR THE FIRST TIME.

I MAY BE THE MOST INTELLEGENT MAN ON EARTH. HOW COOL IS THAT THOUGHT AND POSSIBLE REALITY.

IT APPEARS NO ONE ELSE EVER THOUGHT TO LOOK FOR THE

CAUSE

The

CONTROLLERS

have been looking at earth changes for a long time. They do have at least a rudimental understanding of universal laws of cause and effect.

I operate at a level or plane higher than their understanding but they are close or they are not telling us the whole story of their scientific information concerning co2 becoming carbonic acid in mammal's bodies.

I wonder if whales have

"Osteophytes of the Vertebral Column"

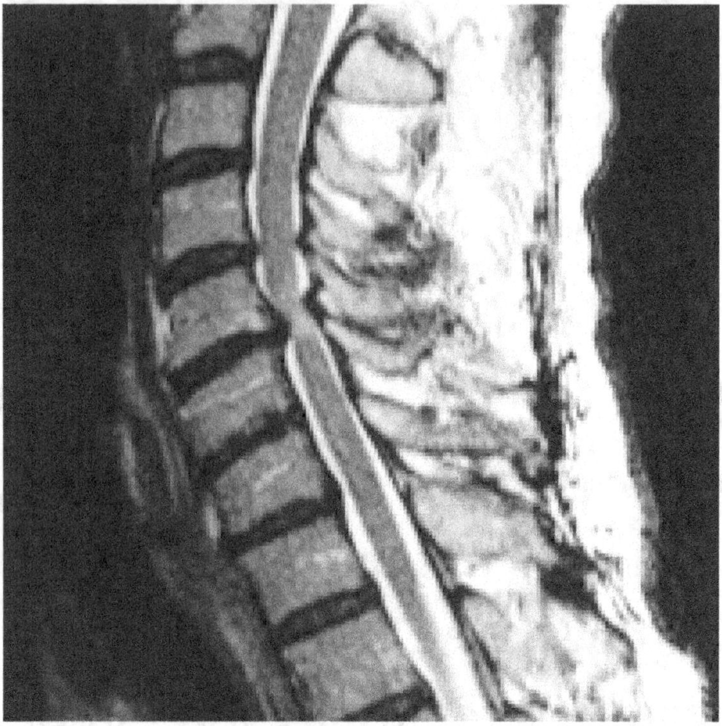

This was me, dying from carbon deposits with a tumor pressing down pushing me to paralysis. A life friend, Dottie Oliver directed me to underground THC squat in Amsterdam and in weeks using gauze soaked in THC resin, my energy began to return within minutes of the first dose of natural magic concentrates in a tincture.

My beyond wanting to live pain was livable.

Within a week more feeling returned to my right side and on Christmas eve 2009 I flew to India to meet a spine surgeon I met through a late night wrong number call to India. It was all a miracle.

To offer information and a course of action we have set up our web information site

PUBLIC EYES MEDIA.ORG

Get a hair sample and Act

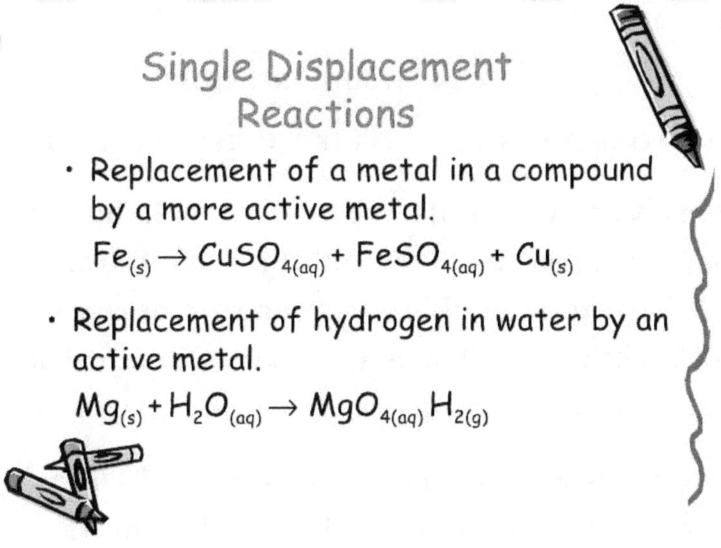

Single Displacement Reactions
• Replacement of a metal in a compound by a more active metal.
$Fe_{(s)} \rightarrow CuSO_{4(aq)} + FeSO_{4(aq)} + Cu_{(s)}$
• Replacement of hydrogen in water by an active metal.
$Mg_{(s)} + H_2O_{(aq)} \rightarrow MgO_{4(aq)} H_{2(g)}$

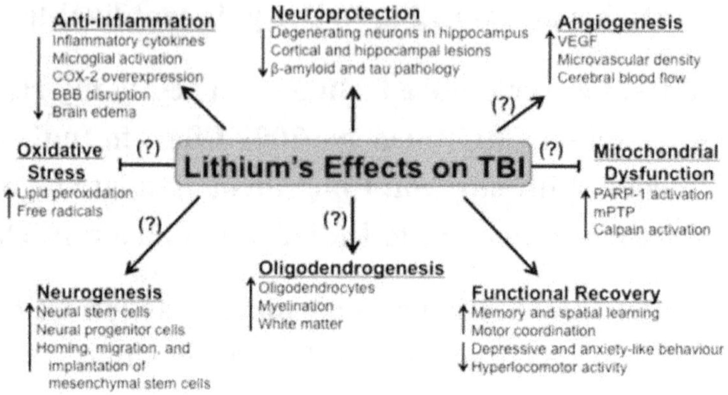

GLOBAL CARBON DISEASE

SMOKY TRUTH

VIDEO PRODUCED IN 2010

NO ONE WAS READY FOR IT 2019 RELEASE PENDING

This is an excerpt of a study that should have been of great concern at the time of release. But in 1961-62 the general thinking was nonexistent concerning global co2 levels that concentrate in us as calcium carbonate.

Or, it was well understood, in a small circle, and the choice was to not awaken the world to the danger and potentially change our way of powering the

world and change the in-power structure of the planet system.

That may be the answer to the ignorance of all thinkers of science.

I also assert that co2 is the brain glue onto which transmuted minerals deposit and steal our minds. I witnessed, as a cameraman, on a video shoot about a man with Alzheimer's who had not spoken to his wife in 10 years because of the disease.

He was put into a hyperbaric dive chamber used to over-come the bends. He had been given a 40 grams of liquid vitamin c intravenous chelation needed to pull heavy metals out of his blood stream. All humans carry at least 17 toxic metals in our blood always.

All metals have disease causing agents especially aluminum that is known as the primary cause of Alzheimer's but few doctors who treat this disease use toxics detection like hair sample, for aluminum and other metals.

One reason: A doctors who orders a hair toxicity test using Medicare Insurance for a routine look, at possible causes of disease in patients over 65, will

lose license and will receive a prison sentence.

Most Alzheimer's patients are on Medicare. (look it up). Maybe 'some-ones' know this game changing evidence and have decided to continue the very successful game of fooling the fluoride fooled.

The universe is powered by cause and effect. Simple and profound. The only conclusion reachable is

solidified **co2** is building in all mammals and will affect all brains. In 1921 co2 levels shot up in

volume.

And correspondingly deposits of calcium carbonate began showing up in human spine.

It is so simple.

Here is 'effect' evidence of co2 concentrated carbonic acid then calcium solids

Carbon solidification began in 1921 the year our Earth reached co2 carrying capacity

A STUDY BY ISRAELI DOCTORS/SCIENTIST

An Anatomical Study of Their Development According to Age, Race, and Sex with Considerations as to Their Etiology and Significance

Hilel Nathan M.D.[1]

[1] **Department of Anatomy, The Hebrew University-Hadassah Medical School, Jerusalem**

"A series of 400 vertebral columns of whites and Negroes of both sexes and of various ages were examined for the presence of osteophytes.

These were classified according to four degrees of development. Osteophytes were first found in the twenties, and the proportion of affected spines increased directly thereafter. In the forties, 100 per cent of skeletons showed first-degree osteophytes.

The other degrees of osteophytes were found in 100 per cent of skeletons of people who were over eighty years of age. The distribution of the osteophytes in the different regions of the spine, as well as their localization on each vertebral body, was found to follow characteristic patterns. The most outstanding features of these patterns were:

1. **The incidence of osteophytes is greater on the anterior aspect than on the posterior aspect of the vertebral body**

2. **Anterior or posterior osteophytes tend to develop more in the concavities of the normal vertebral column or within the concavities of scoliosis or kyphosis.**

3. **Peaks of regional distribution are seen and are related to the normal curvatures of the vertebral column and to the line of gravity crossing them. The presence of osteophytes on the superior or inferior borders of the vertebrae is also related to the spinal curvatures.**

These findings indicate that osteophytes tend to appear more where pressure is greatest. This leads to the concept that osteophytes develop as a defense mechanism in response to pressure. Further support for this theory is provided by the fact that osteophytes are composed of more compact, stronger bone

than the rest of the vertebral body and by the fact that the form and position of the osteophytes on the vertebral body resemble the capitals and bases of pillars designed by architects to increase the resistance of these pillars to compression.

The thoracic spine is characterized especially by the predominance of osteophytes on the right side, a distribution caused by the aorta which runs down on the left. This general pattern of development of osteophytes is similar in all the race and sex groups examined. However, some differences are found: In anterior osteophytes, whites of both sexes show a greater incidence than Negroes, but the difference is not significant; whereas males of both races show a greater prevalence than females, the difference being statistically significant. Regarding posterior osteophytes, a significantly higher incidence is found in whites of both sexes; the slightly higher incidence in males than in females of both races were found to be non-significant.

The soft tissues into which osteophytes may grow and the different pathological conditions which may be produced by osteophytes pressing on viscera related to the vertebral column are reviewed”.

CO2 GOES INTO SOLUTION IN US AFTER BREATHING

co2 SATURATED AIR.

IT THEN BECOMES CARBONIC ACID THEN SOLID CALCIUM CARBONATE THAT

DEPOSITS IN ALL OF US

YES, AND HERE IS MY EVIDENCE. I AM AFFLICTED BY HYPO-CALCEMIA. THIS CONDITION IS CONSIDERED A FEMALE DISEASE.

MEN NOW HAVE THIS PROBLEM AS SHOWN IN HAIR SAMPLES.

THE CAUSE IS TOO MUCH CO2 IN AIR SUPPLY AND LOW EXPLUSION LEVELS

OF THE GAS IN OUR EXHALING OUR AIR INTAKE.

ALUMINUM DISPLACES

CALCIUM. ALUMINUM AFFECTS VITAMIN D ASSIMILATION

ANOTHER FACTOR IS LOW VITAMIN D AND IODINE

WITHOUT THE D-HORMONE CALCIUM WILL NOT BE USED BY THE BODY.

VITAMIN D IS MAYBE AN INTERNAL FIRE STARTER

IODINE IS IODINE

I AM NOT A CHEMIST

DO YOUR OWN STUDY

THAT MEANS EVIDENCE

MUST BE

AVAILABLE IF ONE LOOKS

THE TOP TWO LINES OF THE LOWER PANEL OF MY HAIR TEST SHOW CALCIUM AND MAGNESIUM FLOWING IN MY BLOOD STREAM AND LEAVING MY BODY BY HAIR EXCRETION

WHAT HAPPENS TO OUR HEALTH WHEN WE ARE

NORMALLY TOXIC?

MY RECENT SPINE SURGEON THINKS I MAY HAVE DISCOVERED THE SOURCE OF MOST MASS

DISEASE IN THE MODERN WORLD SINCE 1950.

MY HAIR SAMPLE

I WAS TOLD WAS NOT THAT BAD. WHAT IS BAD WHEN THIS TOXICITY LEADS TO SOME DISEASE!

HAIR ELEMENTS

DOCTOR'S DATA INC

LAB#: H101227-2652-1
PATIENT: Gary Burris
ID: BURRIS-G-00023
SEX: Male
AGE: 60

CLIENT#: 24237
DOCTOR: Anna Davis, MD
Direct Laboratory Services
4040 Florida St Ste 202
Mandeville , LA 70448 USA

POTENTIALLY TOXIC ELEMENTS

TOXIC ELEMENTS	RESULT µg/g	REFERENCE RANGE	PERCENTILE 68th 95th
Aluminum	3.8	< 7.0	
Antimony	0.031	< 0.066	
Arsenic	0.063	< 0.080	
Barium	1.6	< 1.0	
Beryllium	< 0.01	< 0.020	
Bismuth	0.10	< 2.0	
Cadmium	0.038	< 0.065	
Lead	1.4	< 0.80	
Mercury	0.47	< 0.80	
Platinum	< 0.003	< 0.005	
Thallium	< 0.001	< 0.002	
Thorium	< 0.001	< 0.002	
Uranium	0.007	< 0.060	
Nickel	0.23	< 0.20	
Silver	0.09	< 0.08	
Tin	0.17	< 0.30	
Titanium	0.68	< 0.60	
Total Toxic Representation			

ESSENTIAL AND OTHER ELEMENTS

ELEMENTS	RESULT µg/g	REFERENCE RANGE	PERCENTILE 2.5th 16th 50th 84th 97.5th
Calcium	1260	200- 750	
Magnesium	81	25- 75	
Sodium	35	20- 180	
Potassium	17	9- 80	
Copper	14	11- 30	
Zinc	180	130- 200	
Manganese	0.29	0.08- 0.50	
Chromium	0.41	0.40- 0.70	
Vanadium	0.056	0.018- 0.065	
Molybdenum	0.015	0.025- 0.060	
Boron	1.6	0.40- 3.0	
Iodine	1.1	0.25- 1.8	
Lithium	0.012	0.007- 0.020	
Phosphorus	186	150- 220	
Selenium	0.83	0.70- 1.2	
Strontium	3.4	0.30- 3.5	
Sulfur	44900	44000- 50000	
Cobalt	0.016	0.004- 0.020	
Iron	12	7.0- 16	
Germanium	0.025	0.030- 0.040	
Rubidium	0.012	0.011- 0.12	
Zirconium	0.39	0.020- 0.44	

SPECIMEN DATA

COMMENTS:

Date Collected:	12/21/2010	Sample Size:	0.105 g
Date Received:	12/27/2010	Sample Type:	Head
Date Completed:	12/30/2010	Hair Color:	Gray
Client Reference:		Treatment:	
Methodology:	ICP-MS	Shampoo:	

V010.08

RATIOS

ELEMENTS	RATIOS	EXPECTED RANGE
Ca/Mg	15.6	4- 30
Ca/P	6.77	0.8- 8
Na/K	2.06	0.5- 10
Zn/Cu	12.9	4- 20
Zn/Cd	> 999	> 800

©DOCTOR'S DATA, INC. • ADDRESS: 3755 Illinois Avenue, St. Charles, IL 60174-2420 • CLIA ID NO: 14D0646470 • MEDICARE PROVIDER NO: 148453

I recently learned more about **Vitamin D** *deficiency. My calcium/magnesium loss was in part from my hormonal deficiency of the ultra-essential D-hormonal*

Too cloudy to recharge every day!

HAIR ELEMENTS REPORT INTRODUCTION

Hair is an excretory tissue for essential, nonessential and potentially toxic elements. In general, the amount of an element that is irreversibly incorporated into growing hair is proportional to the level of the element in other body tissues.

Therefore, hair elements analysis provides an indirect screening test for physiological excess, deficiency or maldistribution of elements in the body. Clinical research indicates that hair levels of specific elements, particularly potentially toxic elements such as cadmium, mercury, lead and arsenic, are highly correlated with pathological disorders. For such elements, levels in hair may be more indicative of body stores than the levels in blood and urine.

All screening tests have limitations that must be taken into consideration. The correlation between hair element levels and physiological disorders is determined by numerous factors. Individual variability and compensatory mechanisms are major factors that affect the relationship between the distribution of elements in hair and symptoms and pathological conditions.

It is also very important to keep in mind that scalp hair is vulnerable to external contamination of elements by exposure to hair treatments and products. Likewise, some hair treatments (e.g. permanent solutions, dyes, and bleach) can strip hair of endogenously acquired elements and result in false low values. Careful consideration of the limitations must be made in the interpretation of results of hair analysis. The data provided should be considered in conjunction with symptomology, diet analysis, occupation and lifestyle, physical examination and the results of other analytical laboratory tests.

Caution: The contents of this report are not intended to be diagnostic and the physicianusing this information is cautioned against treatment based solely on the results of this screening test. For example, copper supplementation based upon a result of low hair copper is contraindicated in patients afflicted with Wilson's Disease.

Lead High

This individual's hair Lead (Pb) level is considered to be moderately elevated. Generally, hair isa good indicator of exposure to Pb. However, elevated levels of Pb in head hair can be an artifact of hair darkening agents, or dyes, e.g. lead acetate. Although these agents can cause exogenous contamination some transdermal absorption does occur.

Pb has neurotoxic and nephrotoxic effects in humans as well as interfering with heme biosynthesis.Pb may also affect the body's ability to utilize the essential elements calcium, magnesium, and zinc.At moderate levels of body burden, Pb may have adverse effects on memory, cognitive function, nerve conduction, and metabolism of vitamin D. Children with hair Pb levels greater than 1 µg/g have been reported to have a higher incidence of hyperactivity than those with less than 1 µg/g. Children withhair Pb levels above 3 µg/g have been reported to have more learning problems than those with less than 3 µg/g. Detoxification therapy by means of chelation results in transient increases in hair lead.Eventually, the hair Pb level will normalize after detoxification is complete.

Symptoms associated with excess Pb are somewhat nonspecific, but include: anemia, headaches, fatigue, weight loss, cognitive dysfunction and decreased coordination.

Sources of exposure to Pb include: welding, old leaded paint (chips/dust), drinking water, some

I took action! I got on oral chelation to pull the toxins out. I aggressively re-mineralized to replace the essential minerals being displaced daily by toxins in the air, water, and our industrial food

1950

"AGAIN" LIKE FOREST GUMP SAID

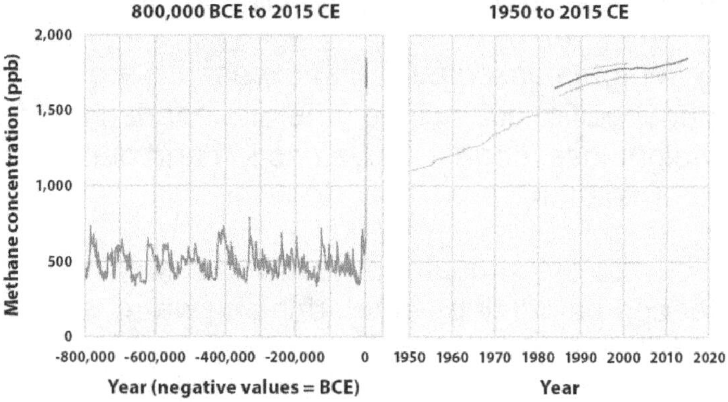

Doctor Dodd

Patient name is Sambo.

He is a mammal

Sambo was very sick when he came to the clinic. The doctor knew something was wrong with toxicity. Sambo had been to several other doctors and now could find out what was killing him. Then he met Doctor Dodd and a world of reality and health restoration was revealed.

Doctor Dodd order a hair sample analysis: To explain read the following

Hair analysis can detect and measure the content of heavy metals and minerals of the hair. This applies to animals and people. Abnormal values of both relate to the same pathology expressed in man and animals.

The Global Environmental Monitoring System (GEMS) of the United Nations Environment Program selected human hair as one of the important monitoring materials for worldwide biological monitoring of pollution.

(From the government book, "Toxic Trace Metals In Mammalian Hair and Nails"

by the US Environmental Protection Agency.)

If your hair reveals the presence of heavy metals (arsenic, lead, mercury, aluminum, cadmium, selenium, nickel, copper and iron) this means the tissues and organs of your body contain them too.

The hair is a spillover from what is in the body. Generally heavy metals cannot be detected by blood, or urine tests.

Heavy metals are toxic, the severity of symptoms being related to the absolutes consumed, and the

time factor involved. Large doses consumed can lead to death immediately, smaller amounts over many months or years produce chronic degenerative diseases.

Spark Plugs of Life

Heavy metals chemically relate to mineral content of the body. For example, Aluminum will displace Calcium, magnesium and manganese due to its valence in the atomic table (remember your high school chemistry?)

This produces a deficiency of these three minerals with metabolic dysfunctions. Minerals are the "sparkplugs" of life. They are involved in almost all enzyme reactions within the body. Without enzyme activity, life ceases to exist.

A trace mineral analysis is preventative as well as being useful as a screening tool.

What are some of the symptoms of mineral imbalance? The same is true for the mineral content of the hair: if you have a deficiency of calcium in the hair, this reflects a deficiency of calcium in the body. Blood tests will reveal normal

calcium levels, because there are physiological mechanisms to keep it normal or anyone would die of a titanic seizure and heart attack.

EXPLOSION OF DENTAL PROBLEMS

AND OSTEOPOROSIS

AND NO ONE ASK WHY. UNTIL NOW

The body will give up its calcium from the teeth, or bone to keep that normal calcium blood level, thus producing osteoporosis (with increased frequency of fractures), dental caries, muscle cramping of the skeletal muscles and muscles of the heart.

On the other hand, if your hair shows an elevated calcium level two or three times normal, then your calcium level within the body may be elevated too. This produces a strong tendency for arteriosclerosis in people and arthritic calcium deposits in the joints of people and animals.

HEAVY METAL

AND NOT THAT LOUD MUSIC

Heavy Metals causing mineral imbalance can lead to the following conditions in people/ or animals: ("Toxic Metal in Human Health and Disease".

Doctor Paul Eck and Dr. Larry Wilson,

Eck Institute of Applied Nutrition and Bioenergetics

Depression-Toxicity with Lead, and Aluminum

Hypoglycemia- elevations of sodium to potassium ratios with potassium deficiency prevents blood glucose from crossing the cell membrane and contribute to inflammations and stress. This can be associated with many heavy metals.

Aluminum toxicity lowers manganese levels that influence blood sugar disturbances

Hyperactivity- can be related to Lead poisoning in children, and some forms of mental illness

Headaches- almost all heavy metals Hypertension- can be due to elevated sodium to magnesium ratio, and elevated sodium to potassium ratios causing edema and water retention in the body with increased blood pressure.

Many heavy metals play a role in this

Arthritis: This disease can be related to Aluminum and Lead's relationship to calcium levels in the body, Copper is implicated in various collagen diseases and rheumatoid arthritis, also elevated Iron's association to Aluminum toxicity ("rusty" joint syndrome) Lowered resistance to infections and healing are related to zinc deficiencies caused by toxic levels of Aluminum and Copper

Hair loss: Selenium toxicity, and premature graying from Aluminum Anemia and bone marrow depression- seen with Iron deficiencies and Aluminum and Copper toxicity

Hypo and Hyperthyroidism- Aluminum toxicity

Prostate disorders: Aluminum toxicity interferes with zinc metabolism necessary for the health of the prostate.

Digestive Disorders- Aluminum can cause muscle cramping and colic,

Copper causes liver degeneration.

Aluminum causes hemosiderosis (a liver disease with increased reserves of iron in the liver) by preventing **metabolism** of Iron.

Diabetes Mellitus- can be caused by a deficiency of Chromium, magnesium, manganese and Vitamin B6

Muscular-Skeletal Disorders - Lead toxicity is linked to **Multiple Sclerosis,** and Aluminum is associated with muscle weakness by its lowering of manganese, magnesium and calcium

Cardiovascular Disease- can be associated with Aluminum, Arsenic, Selenium,

Skin rashes, alopecia and allergies can be due to intoxication with Aluminum, and Selenium,

Mental Illness - high levels of Copper and Iron can cause Schizophrenia, Paranoia, and suicidal tendencies.

Epilepsy- is related to magnesium deficiency- so too is leukemia, kidney and heart problems.

Aluminum displaces magnesium.

Cancer - Beryllium, Chromium, Lead, Nickel, Aluminum, Selenium and Arsenic poisoning can cause cancer

Fetal deaths- mercury toxicity in the mother's system can kill her unborn baby. "20-20" in the recent past aired, stated that, approximately 60,000 fetuses will develop methyl mercury toxicity in-utero this year. Because the mothers eat swordfish, shark and tuna fish.

Cadmium, and Nickel cause congenital abnormalities and pregnancy toxemia. MD's tell us nothing, The FDA is impotent, and our government fails to confront the fishing and corporate industries to do something about the discharges from polluted waters going into the ocean, and contaminated landfills leaching into soils and underground water reservoirs. Kidney dysfunctions, and Cystitis with crystal and stone formation- can be due to Aluminum toxicity of the liver and pancreas

Who Needs a Hair Analysis?

Those Darwinians' who understand

Any human or animal that is found ill and no explanation can be found, or treatment is ineffective.

Hair analysis opens a whole new vista for solving the mystery of illness. An imbalance of minerals can be caused by stress, improper diet, taking the wrong vitamins or mineral supplements or excessive amounts, medications and accumulation of toxic metals from the environment.

It is the latter I wish to discuss.

In my clinical practice, since 1985, when I began running hair analysis on every sick animal (no matter what the cause or symptom of the illness) that came to me, every report showed abnormal mineral content and toxic levels of Aluminum. Many showed various combinations of all the other heavy metals tested.

Every cat showed toxic levels of Mercury (due to high fish meal based cat foods) and Aluminum. Aluminum, Lead and Mercury go to the brain and the nervous system, and thus touches every organ in the body.

Heavy metals are contaminating our food, water and health supplements for animals and man.

The major source is chemical pollution by agribusiness, industry and the public dumping of toxic household wastes in landfills. All these metals upset the mineral balance in the body, which exacerbates ill health by nutritional imbalances.

Only a hair analysis can reveal this, not a blood test, urinalysis, x-rays, MRI, or CAT scan of the brain.

Dr. Dodd's comment: I truly believe man's pollution and his proliferation of noxious energy fields on this planet are at the bottom of every illness and violent act in our society.

Sambo's" Hair Analysis Reports

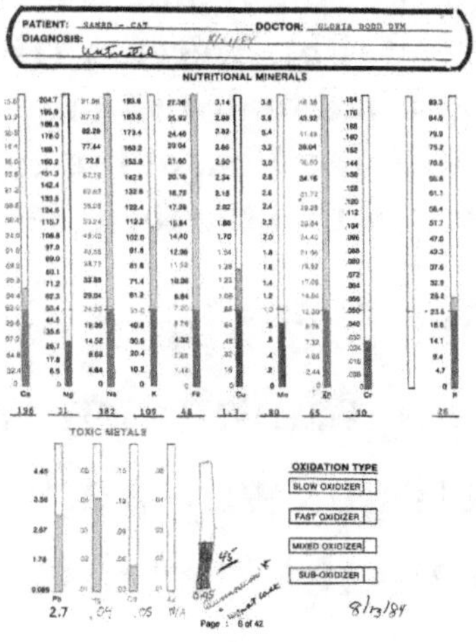

Pretreatment hair sample on 8/24/84: Interpretation: The top of the chart shows the nutritional mineral content of the body, the black bars across each column bar indicates normal levels, any white on the graph below this line indicates a deficiency of the mineral.

The green part of the column above this line indicates an excess of this mineral. Sambo showed an excess of calcium, (Ca) that contributes to osteoarthritis; deficiency of magnesium (Mg), elevated sodium (Na) and potassium (K), increased amounts with secondary toxicity of iron (Fe) and copper (Cu); deficiency of manganese (Mn), Zinc (Zn) loss from the body and deficiency of chromium (Cr).

The lower part of the report shows Toxic Metals: content of Lead (Pb), Mercury (Hg) Cadmium (Cd) Arsenic (As). I had to draw in my own bar, after conversing with Dr. Eck, of the analyzed Aluminum because it was not on the form sent.

(labs were forbidden by secrecy acts from open aluminum testing, a few sampled anyway and sidestepped the criminal federal government. I wonder why?)

The bottom of each bar indicates man's (approximate 180-pound body weight) tolerable level of that toxic metal.

Compared to Sambo's values (at 11 pounds' body weight comparison).

Tolerable man's level of Lead is .089 mg%, Sambo showed 2.7 mg%! Over 27 times more.

Tolerable man's level of Mercury is .01mg%, Sambo showed .04 mg% - 4 times more.

Tolerable man's level of Cadmium .03- Sambo had .05 Tolerable man's level of Aluminum (as determined by Dr. Eck) is .05mg%- Sambo had 45 mg%- 1000 times more than tolerable for man!!!! Dr. Eck was amazed he was still alive.

He told me he had never seen such a high level in man or an animal.

Sadly, 22 years later we see the vastly increased levels of aluminum in our dogs, cats and horses' and human babies' hair samples.

Note imbalanced nutritional minerals in Sambo's graph: lowered Magnesium, Manganese and Chromium, with toxic elevated levels of Zinc, Iron,

and Copper. Calcium, Phosphorous, Sodium and Potassium were higher than normal also.

Six weeks of homeopathic Detox treatments cleansed Sambo's system of the heavy metals. A monitor hair sample was taken. These are the results:

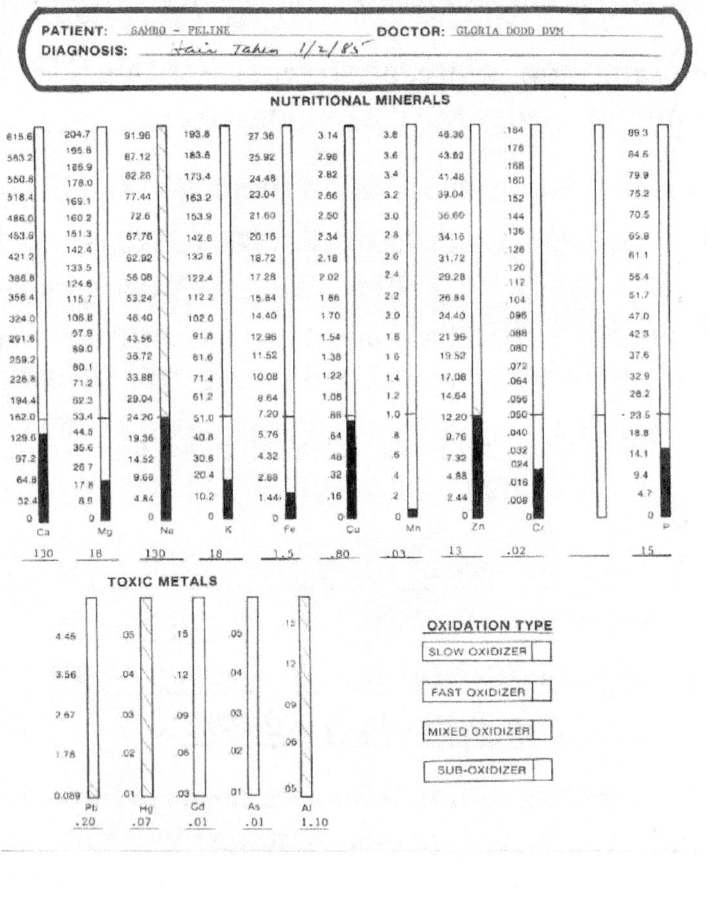

Clinical improvement, and normal blood test values matched the hair results in the lowering of all the heavy metals: Aluminum from 45 mg% to 1.10 mg%;

Lead from 2.7mg% to .20mg%

Cadmium from .05 to .01 mg%

Mercury showed an increase from .04 to .07 mg%: This which meant the tissues were giving up the Mercury to the hair. I continued to use higher potencies of the homeopathic heavy metal detox until all were eliminated. Vitamin Mineral supplementation provided the deficient minerals until all metals were cleared from his system.

Sambo lived for many years.

With proper DETOX care, and MINERAL therapy, humans can clean THEIR bodies and heal. Healthy children and dogs thrive.

THE CHEMICALLY DRIVEN DEATH

OF DOTTIE OLIVER, A LIFE LONG FRIEND

THIS IS A CAUSE AND EFFECT STORY OF BP HORIZON TOXINS KILLING LONG AFTER THE MEDIA LEFT TOWN

HER, AND OTHER DARWINS, STORY OF CANCER HELD AT BAY WITH CANNABIS, IS ONE OF HEALING, QUALITY OF LIFE, AND BUYING TIME IN A TOXIC WORLD.

THE PEOPLE IN THIS BOOK ARE LIVING OUR TOXIC EARTH FOR PROFIT HEALTH STORY: IT IS NOT GOOD BUT THIS INFORMATION WILL INCREASE YOUR QUALITY OF LIFE IF YOU

FOLLOW THIS ADVICE AND GET A HAIR SAMPLE TEST. DISCOVERING THE CAUSE OF DISEASE IS NOT ENCOURAGED.

JUST MORE ADS SELLING YOU ON MORE TOXIC MEDICINE TO COVER ONE THING CAUSING ANOTHER AND ON AND ON TO THE WORLD BANK

TENS OF MILLIONS, ALONG COAST AND UPWINDERS AROUND THE JETSTREAM HAVE SAME LETHAL SECRET. ACTION IS SURPRESSED BY LACK OF MASS AWARENESS EFFORTS.

ALL HEALTH PROBLEMS HAVE A *CAUSE AND CO_2 AND OTHER TOXINS BEING THE CAUSE OF MOST DISEASE INCREASE SINCE 1921 THAT WENT VIRAL IN 1950.*

THIS WAS UNKNOWN TO THE PUBLIC

***UNTIL NOW*.**

GET THIS: A FISHERMAN CRACKED THE BUSINESS GENOCIDE CODE OF PLANNED DISEASE FOR PROFIT

WE ARE IN THE GRASP OF THAT MONOLITHIC SYSTEM PRESIDENT JOHN F. KENNEDY WARNED US ABOUT

TOP GRAPH BELOW

SHOWS 17-TOXICS METALS/CHEMCIALS

THESE WERE CIRCULATING IN HER BLOOD

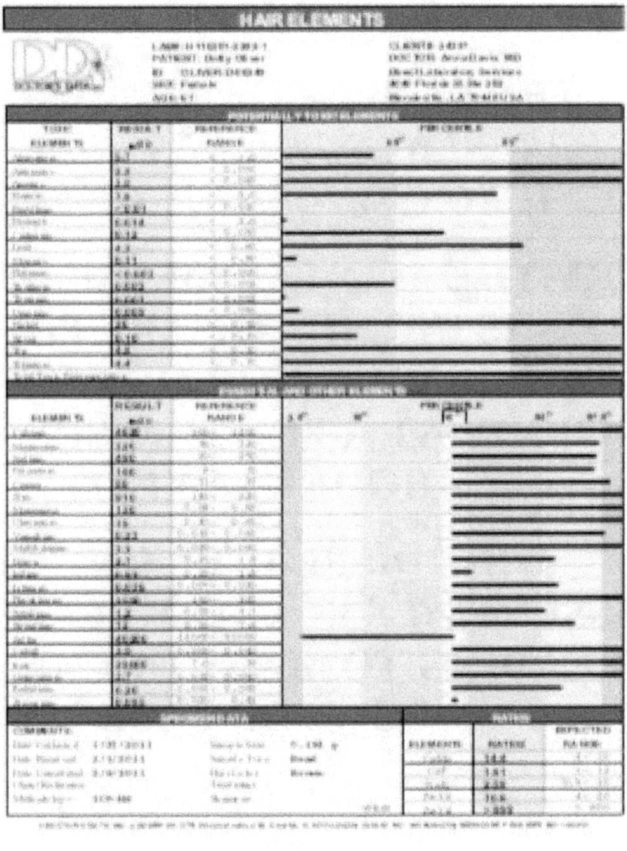
HAIR ELEMENTS

THIS IS THE SIGNITURE OF CANCER

BOTTOM GRAPH

ESSENTIAL MINERALS

THE CENTER OF THE GRAPH REPRESENTS THE BEST AVERAGE ACROSS MILLIONS OF TEST

TRY THE MENTAL IMAGE OF THE TOXINS IN THE TOP ARE PUSHING TO THE RIGHT OR OUT OF THE BODY AT A PROGRESSIVELY HIGHER LEVELS

THIS DISPLACEMENT OF ESSENTIALS IS A MAJOR FACTOR LEADING TO MOST ORGAN AND BONE DISEASE

IF THE LINE IS TO THE LEFT OF CENTER YOU ARE NEGATIVE

\

DEPLETING OR DEPLETED COMPLETELY

LACK OF SULFUR or total depletion WAS A CAUSE OF HER DEATH by lack of immune system fuel: sulfur

HERE IS HOW

MEDICAL-CHEMO AND TOXINS IN OUR LIFE-SUPPORT SYSTEM WEAKENS THROUGH

DISPLACEMENT CHEMISTRY

FIRST LAW OF THE UNIVERSE IS ORDER

UNDERSTAND THE PRINCIPLE OF CAUSE AND EFFECT AND SECRETS IN PLAIN SIGHT WILL BE MADE EVIDENT TO ALL

Is there a cause and effect manifestation of mineral deficiency and accumulative un-necessary toxic burden causing diseases in all man-un-kind?

I intend to reveal a cause and effect relationship of all disease formation of mind and body that may increase the quality of life and give 'Darwin Time' to you and your children.

I intent to raise a reasonable doubt of the normal existential contaminations being an accident:

The planned, for profit and cash flow of a set system, poisoning of humanity is intentional. *The controllers are that far gone in their once ivory temple.*

Now "they" are in trouble and are scrambling for time to build more

underground cities and super escape ships stocked to stay at sea for years

Now that is cool fancy ships filled with guilty fools, built to go out into seas with storms that may start and never stop with 100 foot waves the norm. Have a nice day.

NOW The burden of proof of governmental innocence regarding this toxic disease causing evidence, found in all body protein called human hair is the duty of CDC.

The environmental world toxin managers, producers and geo-engineers are now required to explain and refute this sea captain who out thought the body invasion planners.

Keep reading the evidence is in you enlightening and at times over whelming. But this too will pass if you take action! Period, no other way.

The evidence is now here, in one place, and must be installed in the minds of all the literate world. Social media, at times lurid but at times enlightening, is a weapon against information isolation. Use it.

Informed public opinion and action

now or many will suffer or die of un-necessary sickness, for slow kill profit and population reduction of all people spectrum except the mentally chosen, chosen by fools with antidotes.

Keep reading we now know the

antidote to the chemical invasion of our bodies and minds.

ACTION THROUGH AWARENESS,

AWARENESS THROUGH EDUCATION

THE METHOD OF DISEASE FOR PROFIT IS DISPLACEMENT OF ESSENTIAL MINERALS LEADING TO ORGAN AND BONE DISEASE

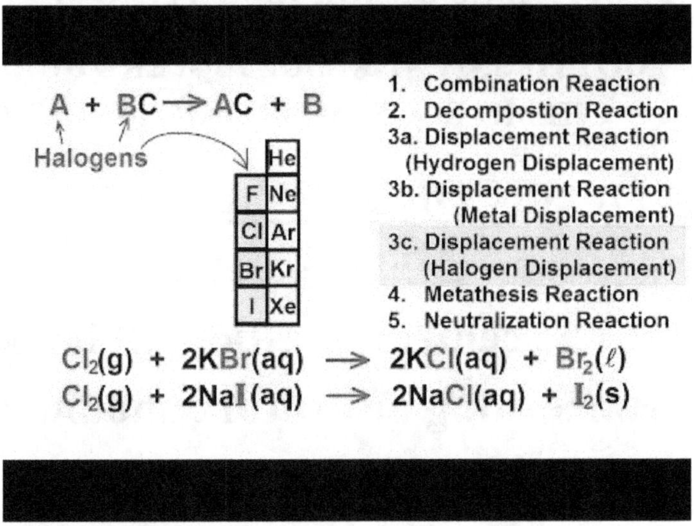

A + BC → AC + B
Halogens
He
F Ne
Cl Ar
Br Kr
I Xe
1. Combination Reaction
2. Decompostion Reaction
3a. Displacement Reaction (Hydrogen Displacement)
3b. Displacement Reaction (Metal Displacement)
3c. Displacement Reaction (Halogen Displacement)
4. Metathesis Reaction
5. Neutralization Reaction
Cl2(g) + 2KBr(aq) → 2KCl(aq) + Br2(ℓ)
Cl2(g) + 2NaI(aq) → 2NaCl(aq) + I2(s)

PUBLISHER DOTTIE OLIVER'S HAIR SAMPLE SHOW ALMOST COMPLETE ESSENTIAL MINERAL DISPLACEMENT. CAUSE OR EFFECT OF DISEASE; I AM NOT SURE. BUT, HALOGENS IN YOUR DRINKING WATER, BY LAW, ARE PART OF THE REACTIVE CAUSATIVE AGENTS OF DISEASE PROCESS.

GIVEN THIS KNOWLEDGE MY CONCLUSION: THE STATE AND FEDERAL GOVERNMENTS UNDER DIRECTION OF

DEEP STATE MANAGERS, AND HANDLERS, ARE REDUCING OUR NUMBERS IN A SLOW PROFITABLE DIVIDENDS PAYING MANNER.

DOTTIES' TEST RESULTS WERE DURING CHEMO EUGENICS TOXIC LOADING

SECONDARY DISEASE ASSURANCE?

The central question all books concerning our future should ask:

Does the potential of saving a nation's OR WORLDS' work force, by pre-emptive medicine and diagnostics, not AMA recommended, nor Medicare friendly, seem the logical thing to work for by our congress and medical community?

WHY CENTER FOR DISEASE CONTROL, NOW A THE

MILITARY RULED ORGANIZATION, NOT LOOKING IN RIGHT PLACE USING APPROVED TOOLS, HAIR SAMPLE, FOR CAUSE

The Medicare system will pay for toxic

evaluation, using protein or hair, if the person has experienced a catastrophic and obvious to others toxic contamination.

In other cases a physician will be subject to charges of fraud and felonious conduct if he uses and changes Medicare for hair test to ascertain subtle disease symptoms.

THEY KNOW AND THIS EVIDENCE IS REAL. WE CAN NOT BLAMEALL OUR DOCTORS!

Typical of low level toxins, when combined, are the causative agent in MOST disease of mind and body.

It seems this government knows but do not want to know.

If the people, or you, know the level of assault on the bodies and our children will they react? Will group coalescence happen or will we remain separate?

The answer is NO: for 100's ***of millions of Americans and other white drug slaves around the most world. All major***

fluoride programs are in the weaponized and developed world.

This is be-cause **the aluminum and fluoride put in our bodies via our water and air stop the chemical functions required for reactive thought to a massive implication.**

YES: For Europe and those 'stupid' less developed countries who do not use the mind control water carried drugs willing. They are hit differently. But many there know what is happening and have said to me, "Oh America so wrong!" Yeah, I travel as a Canadian when in aware nations.

An average child of the world system using the systems' water, food and air will have accumulated 35% of the life burden of poison by 5 years of age.

Many at these ages show early signs of carbon poisoning.

The children of the "Deep State" have less than 1% of this burden, through detox and mineralizing all through life till they take over their fifth column post, assuring a longer, mentally clearer life of dominion planning.

The toxins are not JUST locked in fat and tissue they are flowing in your blood and coming out when the body is exuding hair or protein. Sperm and eggs are contaminated.

Women in Central America, at one time, had so much DDT in breast milk the milk could be used as an insecticide on plants.

Recognition of the profit potential from pre-determined disease, due to the combined levels of legal toxins allowed

in our bodies by law as a causative factor is vital awareness. "Nothing is as it seems".

Recently, in 2019, a CDC Center for Disease Control doctor encourages all to take the 74, yes 74 toxic vaccines using the warped logic that "breast milk has more aluminum than vaccines. 74 vaccines in 15 years assure disease process and reduces the life expectancy.

Population reduction for profit, yes, and

I agreed in part with this thinking.

A 1949 movie, 'The Third Man' was about this very thing.

If a controversial idea is in the movies, it is there as a thought program to give plausible deniability to cover the reality of a real event in viewers' minds.

We are convinced that pre-emptive medicine is as profitable as blind medicine, or blinkered medicine.

We are convinced a healthier society yields greater profits to the whole.

Hair it is!

Life Support System

Some feel something is wrong with our life support system

Then wonder why out loud and not hear a clear answer

Then some wonder why the planet's global

weather is changing fast

Then I wonder are we having mini or human level
'internal chemical weather' changes

Then they ask why as a group

But are told stories by the few in charge

of the flow of information

They say it is not real

But your instinct and intellect

know it is

To be puzzled by things around us

In the natural world

That defy rational reasoning

But there is none from our leaders and the others

who do know more than we do

that feeling is the worst kind of mental confusion!

But there is clarity out there

And it comes piece by piece

Brick by brick

Until one day you know the truth

Then you can begin to do your part

When you know, what must be done

WHY!

I became an environmental/human rights activist long before I knew about the chemtrails spraying program.

In my work as an activist I chose to follow a direction with hands on practices that show real results that could help many.

No religious slant just pragmatic efforts to make life better on our planet. There are ways to help protect ones' self from these toxins in our skies; from a toxic world.

We have heard how polluted we are as a people. We have heard of the probability of many of us developing cancer. Being a pessimist who doubts most government data sources I did my own research, and using hair sampling, I considered the matter of toxic burden in the bodies of a few American citizens.

These following results (see following hair toxicity graphs) showed us the body burden of toxic heavy metals in all tested. These tests show close to what you carry, as a time bomb in every cell.

The test is also an indicator of the toxic burden all people and their children and pets carry as toxic time bombs.

Knowing what is making you sick is three quarters of any battle for health. Precise knowledge of the possible cause enables you to focus on the enemy be it lead, mercury, aluminum or any combination of 17 toxics metal that were tested for and that are in all of us. "The average child celebrates a fifth birthday, having received more than 1/3 of his total lifetime exposure to pesticides."

Deep State antidote treated children have 1% my son and grandchildren toxic burden. They are predisposed for a longer life of control. Soon no one to control will be the reality. But when the mission is DNA dominance over people of this planet is the goal then success is imminent.

We all have poison in us from careless and inefficient technologies that we were given; we were given toxic weapons to use on ourselves called cars. Then were told drive all you want and don't worry—dilution will be the solution.

The problem the planet and all of us mini planets

(people) have is simple: we are now saturated, and all that toxin metal pollution is now spilling from all organs as disease. No place is safe from pollution, especially our bodies.

However, cells can be cleaned. Knowledge is the first step. Now that I have seen many cases of heavy metal poisoning and depletion of mineral disease it is easy to see disease in people.

Like all of you I have toxic metal contamination. The difference between you and me is this: I know what made me sick, you are only wondering.

I know from a hair sample what is inside my body that should not be there. You can have the same power knowing gives. There are remedies to many toxic metal poisonings. If you know you can act.

In my search for solutions to chemical poisoning I met Dr. Gene Lenz Of Remedies in Key Largo.

He has helped many chemo poisoned people with his chelation and mineral balancing therapy. He certainly helped me get cleaner and mineral balanced. Then I helped myself by diet change and halogen free water.

After showing the hair samples results that are in this chapter to Dr. Gene I went to see the well-known environmental healer Dr. William Rae in Dallas in 2011.

I wanted to see if what I was finding in my one of a kind citizen initiated toxic contamination research project was significant.

He had used hair sampling in his practice for decades. Over 35,000 samples over 30 years.

After seeing several hair sample results, he saw high levels of Uranium in some of the fishermen who were near the Deepwater Horizon oil. It was highly radioactive or that was the conclusion of some other researchers who chose to remain silent to the press.

Even more astounding were the results of five people in the Florida Keys. All, were over the red line with uranium contamination.

He also said the levels of lithium and iodine were radially variable. This is the cause of the loss of emotional stability. Certain toxins affect the fight or flight response. It appeared turned off and many stayed where they knew the air was toxic.

He was very interested to learn that there are chemicals that are related to the chemical spraying of the atmosphere by the Deep State Airforce on orders from the United Nations.

Agenda 2100 is real, look it up do your own research. Do not believe me blindly do your own research.

The marching order is population reduction of half of the planet in the next 20 years. The toxics designed for profitable kill are in place inside all of us and our animals. Whales samples show 5000 parts per million aluminum in all samples taken over the last 25 years.

If you think this implausible then consider the convictions of Bush 2, Rumsfeld, Cheney in others countries for war crimes and think are these controllers capable of this culling of the population.

YES, is the answer: A mad man named

Edward Teller is the psychopath responsible for killing us as revenge for the holocaust that killed 50 million humans IN A "HOLOCAUSTIC WAR" TO QUOTE EISENHOWER and not 6 million self-

chosen. He convinced the insurance industry to control weather to protect dividends. Look it up "Planet Needs a Sunscreen" Wall Street Journal. His story diverted truth from NASA/NAZA

Do you need better answers to your questions about autism, Alzheimer's, aluminum aerosols in our air and in all mammals' worldwide including you?

Would you like to discuss, in an action/focus group life controlling issues like fluoride in our water, like unexplained anger, like fatigue and severe headaches that come and go with the movement of the winds and clouds of toxics waste from violating industry in your own backyard, as well as other states and countries.

Do you ever feel just sick and sleepy for unexplained reasons? Well this is maybe happening while our air force is flying over our heads spraying x's and streaks and patterns that grow into thin clouds of toxins material that settle on all below.

A group of mothers in North Little Rock Arkansas began social situation meetings in 2003 to discuss life and soon the subject of headaches in all their children was discussed.

The conclusion was: When Germanic Deep State Airforce, that includes commercial logo planes, spray "chemo-head" or group headaches occur among those with growing and absorbing brains.

Yes, the infiltrations of this century in America placed agents into power here. The Corporatists are our government and no regulation that protect the life support system is allowed openly discussed in mass media. PBS does fine work but their omission of important discovery is one of sins of omission

The national education system was among the first controller grabs along with NASA'S FOUNDING. President Johnson, an obvious NAZI IDEA, sympathizer, under the cover of civil rights advocacy while supporting corporate control over the planet and chemical enslavement by proxy.

Johnson was the point man for Von Braun, a Nazi rocket scientist brought to this country after WW2. He pushed funding for this man, an enemy agent and many more like him in other disciplines.

Johnson and NY German decent bankers, snuck these hundreds of "sci-techno elite" into powerful government and academic positions at Universities.

These silent invaders were capable of influencing Presidents the wrong way. Eisenhower warned us of them.

BLOOD EVIDENCE: The mystery of who knows what deepens when nothing was said in the Arkansas and national press when a snow of white powder like material covered many states September, 28th 2010. This was witnessed by many people. The sky was clear blue and it snowed.

See Vitamin D deception on Geo-Engineering Watch on U-tube.

THE IMAGE CIRCLES ARE NOT RAIN

OR A LENS PROBLEM

IT MAY HAVE BEEN PARTICLES OF CARBON SNOW AND ALUMINUM MATERIAL THAT FLOATED DOWN AFTER DAY OF CHEM SPRAYING SEPTEMBER 28. 2010

THE NEXT DAY HORSES AND WILD LIFE WENT MAD FOR DAYS.

THIS SICKENED MANY

AT THIS OUT DOOR CONCERT

BY DAVID KIMBROUGH AND GUESTS

Arkansas was the center of a blizzard of this toxic material. (New evidence indicates this material is "carbonic snow"). Above during that day, as far as one could see, were our air force and private corporate jumbo jet planes spraying something with a distinct smell and nothing was said on the "Fake media" (to quote our crazy President Trump. Images of this can be seen in my video called "Vitamin D Deception".)

To be empowered by the solving of the puzzle or enough of it to see some of the truth about the condition of our environment is freeing and directing.

One-way to see what is unseen and coming into you with every breath and with every drink, if you drink any fluorinated city water or rain fed springs and wells, is a simple hair sample test.

To 'see' the toxins in a hair report and read the symptoms of a heavy metal poison on a body might help you self-diagnose your mysterious illness or fatigue that many doctors might miss or cannot talk about.

To hear there are some ways to remove them from your body without large medical cost is a very **empowering** and **encouraging** feeling.

Solving the mystery of what is making many feel sick and tired can be done in a 'Chem clean detox workshops' that we are offering to communities and any group wanting answers and action and there we will focus, as a group, on detoxifying your community.

This may be the best tool for taking back your part of this living world. And that is your own body and that of your children.

Authorities will say "we all have it is us" and dismiss this evidence as common place and acceptable. And that is because someone is making money letting this stuff fill our life support system with disease.

Know this many diseases can be detected by these toxic chemical markers found in our hair. A lot can be learned with our hair testing program.

The evidence is in our bodies.

It is said that "our eyes are the doorways to the soul." Eye color tells all's evolution story.

Hair color is another story teller. Hair growth is the only natural and universal body exit mechanism for many unnaturally occurring toxic compounds and metals. It is not common knowledge, or part of science, being taught to all students universally.

A hair toxicity screening should be the first test all doctors do but this is not the case. This almost fool proof test is not taught in all medical schools or offered by allopathic medicine.

First step to better health and disease recovery is to discover and expose the cause of a disease of mind and body which is often toxic metal and or displacement of essential minerals

Many of these diseases causing toxic metals and displacement of essential minerals show up with hair testing.

Then Remove the cause and disease will go

away correspondingly and become less likely to re-invade your body. It is very simple—eliminate the cause of disease, toxic chemicals and the need for continuing treatment for a long list of diseases will be lessened. We would not need doctors nearly as much.

The cure; Immediately re-mineralize yourself, family and dogs.

Ever second our blood carries these mineral building blocks and healing blocks of mineral to cells that need this to stay stable.

Not a once a week deal, every day you must do this as organically as possible.

Our industrial food is bio-

engineered and does not supply a proper minimum of essential minerals and other trace balancers essential for daily cell recovery and nourishment. Every second our blood, like a mini red river feeds our greater internal sea.

Our body must have these minerals, in balance, to repair our cells every second. The oceans of the world are sickened massively from too many essential nutrients.

As above, So below.

A tide has turned and now harmful bacteria has overtaken hundreds of miles of populated downstream coast.

This same process of too many nutrients is why so many obese have so many diseases.

Cause and effect in action possible in both directions.

A STAR

Is built of minerals. It will die soon after birth if out of mineral mind or mineral balance.

Our planet was also built of molten minerals and essential metals. All minerals are essential to something.

Our planet can die if we stay out of balance much longer. This is frightening but exhilarating stuff. This maybe evolution in flash action.

This is the information needed to give a unifying principle to all humanity.

We breath the same air.

We are told toxic metals and gases in our life support system are normal and acceptable: the mass human collateral damage of doing business is considered, and touted, as normal by profiteers.

Dying aged is normal not mass sickness of children, dogs and young adults. Given what I have written 'adult' is a skeptical notion at best.

Nothing can be farther from the truth. These toxic chemicals are sickening to everything living. The exception: 'Darwins' and controller minions needed, know there are antidotes. No slaves, no slavery!

Aluminum, mercury, arsenic, uranium, lead and more are present in all of us, and our children. These are all ticking time bombs connected to the cause of most disease and few are natural.

Before 1945, refined aluminum was not in the open environment. Today, on some days, the sky is filled

with enough metallic aerosol material to snow down on us, courtesy of our sovereign, and immune from criminal prosecution, Air Force.

All mothers milk has levels of aluminum that is transferred to fetus and children: This predisposes disease. Period!

The result, of these toxic spreading actions, is a wave of disease and death circling the planet and hitting the developed world and the third world every day. Finally, equality.

Starving in Biafra or emaciated in Intensive care or hospice kept chemically alive. What is the choice that is sane?

<u>'Downwinders'</u>: Since the BP well blow out in 2010 when concentrated oil spill chemical spraying at low and high altitudes, many middle-aged people and very young in Louisiana, Mississippi and Florida have died or are very ill.

Many new born who were down wind of the spill have chronic illnesses in Louisiana and Mississippi. The city of New Orleans was hit hard with spill related toxins and the sky over the city was filled with very low flying chemo trail planes.

William Rae, an internist and alternative care doctor in Dallas and some state researchers say that one great health threat is the wind that brings high concentrations of these airborne poisons to Texas, Arkansas and surrounding states for months on end. The Gulf of Mexico ecosystem took the big hit.

Every year dolphins die of natural cause like humans. 2019 finally anteed up the evidence. Human, sometimes fatal bacterial infection is everywhere along the coast and up the river, and dolphins, show us our reality that is coming or is here and picking up speed.

This central time zone receives high levels of pollution and chemo spray. It come from three direction at different times of the years. From the west comes everything that China, Mexico and California and others states put into to the air.

From the east comes all kinds of pollution from Canada, New England, New York and all other areas. And from the south comes unbelievable amounts of toxins from refineries and chemical plants and over these areas is massive chemical spraying to cover the holes burned into the stratosphere by these rising toxins and co2 and heat.

Often understated is the radioactive dust that blows to the east from New Mexico, Nevada. This dust if contaminated with still radioactive atom bomb waste from stupid bomb test by mad men in New Mexico and Nevada. The material that was forced into the stratosphere where there is no weather to waste in down quickly. Now still "hot"

bomb material is drifting down **Daily**

and will continue for millions of years. Mental ill gambles gambled with our life support system with a flush it away toilet flush mentality.

The evidence is also in our bodies, or it is

not in our bodies. Toxins displace hormones like

Vitamin D. Vitamin D deficiency is linked to inflammatory and long-latency diseases such as rheumatoid arthritis, multiple sclerosis, diabetes, tuberculosis, and various cancers, to name a very few.

How It Works

Hair Elements Report

Hair is an excretory tissue for essential, nonessential and potentially toxic elements. In general, the amount of an element that is irreversibly incorporated into growing hair is proportional to the level of the element in other body tissues.

Therefore, hair elements analysis provides an indirect screening test for physiological excess, decadency or maldistribution of elements in the body.

Clinical research indicates that hair levels of special elements, particularly potentially toxic elements such as cadmium, mercury, lead and arsenic, are highly correlated with pathological disorders.

For such elements, levels in hair may be more indicative of body stores than the levels in blood and urine. All screening tests have limitations that must be taken into consideration. The correlation between hair element levels and physiological disorders is determined by numerous factors.

Individual variability and compensatory mechanisms are major factors that affect the relationship between the distribution of elements in hair and symptoms and pathological conditions. It is also very important to keep in mind that scalp hair is vulnerable to external contamination of elements by exposure to hair treatments and products.

Likewise, some hair treatments (e.g. permanent solutions, dyes, and bleach) can strip hair of endogenously acquired elements and result in false low values. Careful consideration of the limitations must be made in the interpretation of results of hair analysis.

The data provided should be considered in conjunction with symptomology, diet analysis, occupation and lifestyle, physical examination and the results of other analytical laboratory tests.

A LIST OF SYMPTOMS DISEASE

SUFFERED BY MANY LIVING

UNDER THE ALUMINUM EXHAUST FROM THE SHUTTLE AND OTHERS AND CHEM-SPRAYING ACTIVITIES.

I HAVE SEEN

THIS SPRAYING ALL AROUND THE PLANET.

Of maybe equal or greater contribution to body toxicity level leading to disease are the small legal levels of thousands of chemicals in food and everything we touch. "Take nothing for granted" said Eisenhower in 1961.

That means to me someone somewhere knows the results of multiple chemicals and the potential for specific disease. Yeah they that crazy.

- **Headache**
- **Brain fog**
- **Fatigue**
- **Low energy**

- **Compromised immunity**
- **Disorientation**
- **Difficulty paying attention and concentrating**
- **Sinusitis**

- **Skin discomfort/irritation**
- **Joint pain**
- **Muscle pain**
- **Asthmatic (Breathing difficulties)**
- **Dizziness**
- **Insomnia**
- **Memory loss**
- **Eye problems (blurred or fuzzy vision)**
- **Nausea**
- **Liver problems**

Gallbladder dysfunction

Tinnitus (distant ringing in ears or high pitched sound after spraying)

- **Neck pain**
- **Scratchy throat**
- **Allergy symptoms**
- **Hay fever out of season**
- **Flu-like symptoms**
- **Susceptibility to colds**
- **General weakness**
- **Anxiety**
- **Lightheaded or faint**
- **Depression**
- **Coughing**
- **Sneezing**
- **Shortness of breath**
- **Vertigo**

- **Anger/Rage/Frustration issues**
- **MORGELLONS disease**

TRANSCRIPT OF 'VITAMIN D DECEPTION' FILM INTERVEIW

DALLAS DOCTOR

Dr. Karen Asbury

Her work as an internist and alternative care doctor was superior to most other doctors who practiced by the book. She sought better disease diagnosis after understanding toxic loading in all mammals.

"I am a Doctor of Internal Medicine alternative care; "when they come in (patients) they are so sick they can hardly walk through the door—and it is chronic fatigue, fibromyalgia, heavy metal toxicity, environmental illness. And whatever label you want to put on it almost everyone is deficiency in Vitamin D and Iodine".

"Over the past year or two I have tested several hundred people for iodine only four have come

back normal and out of that number maybe 6-8 came back (from the lab) with normal Vitamin D".

"We do not expect this generation of kids now, the ones born after the year 2000, to outlive their parents"

ME: that is already going on in Louisiana!

Doctor: "It (the die-off younger people) is going everywhere, not just here".

Dr. Roby Mitchell MD

Amarillo Texas, "every year we have a health fair (in Amarillo) and check peoples blood work—I been doing that for maybe 8 years or so—maybe 600 people I probably had maybe 3 or 4 people--- you can count on one hand- out of that 600 who have normal vitamin d levels-

Phyllis Carr, Dietician

Arkansas Heart Hospital

"I do direct patient care and I get all their labs and I do diet therapy of course they send them to me when their vitamin d is low—and the doctors say get their vitamin d up, but you can't do it with food"

"every patient that I have seen, and if they have had their Vitamin D levels checked they have been low. It doesn't matter whether they have been in the sun, out of the sun—use sun screen—don't use sun screen—whether they are fair skinned—dark skinned—it doesn't matter—most of them have been very low at levels around 11 or 13 which is severely low and (during many years as a health professional) I have yet to see a patient who is normal."

"Vitamin D is a vitamin that acts like a hormone and controls many of your bodies' functions.

Question? Which is it vitamin or hormone? It is some of both vitamin and hormone. Without Vitamin D, you cannot lay down bone, in fact, low vitamin D can break down bone. D is involved in so many processes that if you don't have it you are going very sick!"

Gene Lentz PhD

Remedies Nutrition in Florida

Me: question: what is the situation with Vitamin D with the people who come here? (to his body detox center in the Florida Keys

"We are finding now about a 90% deficiency in Vitamin D"

Hypoglycemia is when the body has too much calcium. Calcium or calcium carbonate deposits into the spine and joint of almost all people according to a 1960's study from Israel. (Aluminum interferes with calcium levels in the blood)

This photo is what a carbon bone growth looks like before it is cut into. This was my spine and the doctors who saw the MRI have said it is one of the worst cases ever.

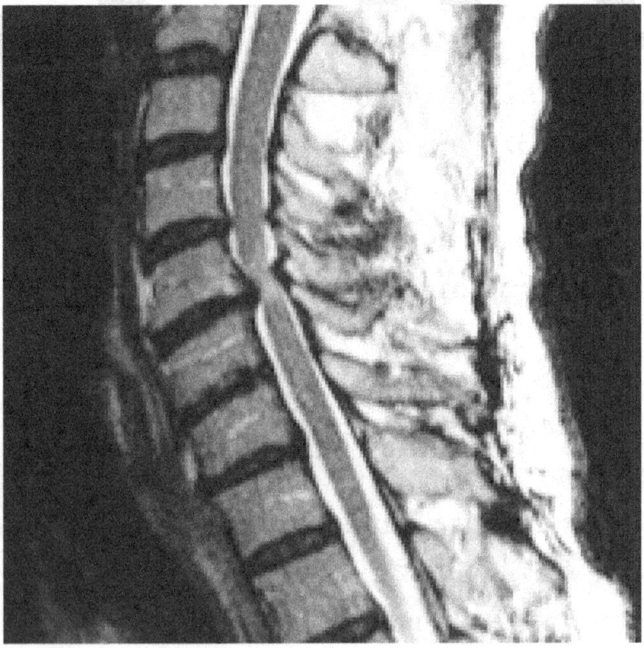

CARBON DEPOSIT FORMED FROM CO2 EXCESS IN AIR.

EXCESS CO2 BECOMES CARBONIC ACID AND THIS IS A FACTOR IN ALZHEIMERS DISEASE WHICH IS CURABLE IF DETOXED, IN TIME, FROM EXCESS CO2 "GLUE' DEPOSITS

This GLOBAL CARBON DISEASE is what many, or most, of you will be experiencing in the future. Your children will experience this, if they do not die from other toxic caused disease because this

'Global Carbon Disease'

is in all of us.

My carbon disease deposits hurt terribly and grew for many years before the pain and the creeping paralysis almost took my life twice plus a near death experience in recovery.

I have lived co2 depositions in the starkest terms and it is no fun but will build strength of character if you persist and work on, but not until almost too late like me, and use the wonderful surgery available.

I was told by Dr. Arvind Kulkarni my surgeon in Bombay/Mumbai India, that I was lucky to make it into his office. "You could have been a vegetable by stepping high off the curb". He is considered by many to be the India's best spinal surgeon. Saved my life! Ten years later another master of carbon deposit removal, Dr. Richard Peek, saved my ass again. During my examination with him, had to wait in line but he is worth the wait, He said twice he had "quadriplegics with less compression. He could not understand why until I told him I was using an energy concentrating cannabis extracts. And I had been opioid free for years by then using cannabis correctly. I was on a long list that quickly went from months to a week. I walked into the hospital knowing I could not walk out again.

Dr. Peek is a master of giving confidence that he would do his best. I like people like that who, like me, will always do their best.

After seeing my mineral displacement hair sample evidence and when I ask if this was the cause of all this disease. He said, 'Yes".

This carbon build up in our spines and other areas cannot be stopped it is a constant intake with every breath. The impact can be lessoned with diet, chelation and hyperbaric treatments. The controllers do this regularly.

Cannabis is essential because it is the delivery material of the planet for extreme high vibration trace minerals or essential brain fuel.

"WE ALL LIVE DOWNSTREAM"

Madness BP style

In many spill workers and those living within 500 miles of spill, those downwind for

months, were shown to lose lithium, a mind chemistry balancing mineral.

Maybe the loss of this mineral is the reason, along with co2 toxicity, for so much madness around the planet.

Lithium

the chemical element of atomic number 3 is a soft silver-white metal. It is the lightest of the alkali metals. lithium carbonate or another lithium salt, used as a mood-stabilizing drug.

Lithium Low

"Lithium (Li) is normally found in hair at very low levels. Hair Li correlates with high dosage of Li carbonate in patients treated for Affective Disorders. However, the clinical significance of low hair Li levels is not certain at this time.

Thus, hair Li is measured primarily for research purposes. Anecdotally, clinical feedback to DDI consultants suggests that low level Li supplementation may have some beneficial effects in patients with behavioral/emotional disorders.

Li occurs almost universally in water and in the diet; excess Li is rapidly excreted in urine.

Li at low levels may have essential functions in humans. Intracellularly, Li inhibits the conversion of phosphorylated inositol to free inositol. In the nervous system, this moderates neuronal excitability. Li also influences monamine neurotransmitter concentrations at the synapse (this function is increased when Li is used therapeutically for mania or bipolar illness).

Below is the test results of a woman who seems very sad all the time. This showed us the reason for her situation. Natural remedies, not prescribed medication, was the best course of actions

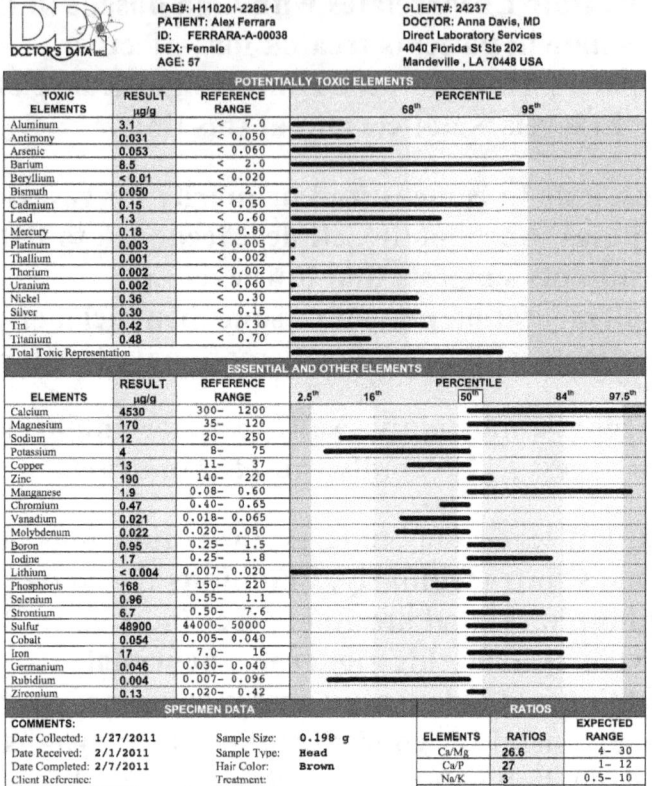

HAIR ELEMENTS

DOCTOR'S DATA INC.

LAB#: H110201-2289-1
PATIENT: Alex Ferrara
ID: FERRARA-A-00038
SEX: Female
AGE: 57

CLIENT#: 24237
DOCTOR: Anna Davis, MD
Direct Laboratory Services
4040 Florida St Ste 202
Mandeville , LA 70448 USA

POTENTIALLY TOXIC ELEMENTS

TOXIC ELEMENTS	RESULT µg/g	REFERENCE RANGE	PERCENTILE 68th 95th
Aluminum	3.1	< 7.0	
Antimony	0.031	< 0.050	
Arsenic	0.053	< 0.060	
Barium	8.5	< 2.0	
Beryllium	< 0.01	< 0.020	
Bismuth	0.050	< 2.0	
Cadmium	0.15	< 0.050	
Lead	1.3	< 0.60	
Mercury	0.18	< 0.80	
Platinum	0.003	< 0.005	
Thallium	0.001	< 0.002	
Thorium	0.002	< 0.002	
Uranium	0.002	< 0.060	
Nickel	0.36	< 0.30	
Silver	0.30	< 0.15	
Tin	0.42	< 0.30	
Titanium	0.48	< 0.70	
Total Toxic Representation			

ESSENTIAL AND OTHER ELEMENTS

ELEMENTS	RESULT µg/g	REFERENCE RANGE	PERCENTILE 2.5th 16th 50th 84th 97.5th
Calcium	4530	300- 1200	
Magnesium	170	35- 120	
Sodium	12	20- 250	
Potassium	4	8- 75	
Copper	13	11- 37	
Zinc	190	140- 220	
Manganese	1.9	0.08- 0.60	
Chromium	0.47	0.40- 0.65	
Vanadium	0.021	0.018- 0.065	
Molybdenum	0.022	0.020- 0.050	
Boron	0.95	0.25- 1.5	
Iodine	1.7	0.25- 1.8	
Lithium	< 0.004	0.007- 0.020	
Phosphorus	168	150- 220	
Selenium	0.96	0.55- 1.1	
Strontium	6.7	0.50- 7.6	
Sulfur	48900	44000- 50000	
Cobalt	0.054	0.005- 0.040	
Iron	17	7.0- 16	
Germanium	0.046	0.030- 0.040	
Rubidium	0.004	0.007- 0.096	
Zirconium	0.13	0.020- 0.42	

SPECIMEN DATA

COMMENTS:

Date Collected:	1/27/2011	Sample Size:	0.198 g
Date Received:	2/1/2011	Sample Type:	Head
Date Completed:	2/7/2011	Hair Color:	Brown
Client Reference:		Treatment:	
Methodology:	ICP-MS	Shampoo:	

V010.08

RATIOS

ELEMENTS	RATIOS	EXPECTED RANGE
Ca/Mg	26.6	4- 30
Ca/P	27	1- 12
Na/K	3	0.5- 10
Zn/Cu	14.6	4- 20
Zn/Cd	> 999	> 800

©DOCTOR'S DATA, INC. • ADDRESS: 3755 Illinois Avenue, St. Charles, IL 60174-2420 • CLIA ID NO: 14D0646470 • MEDICARE PROVIDER NO: 148453

"It's easier to ask forgiveness than permission".

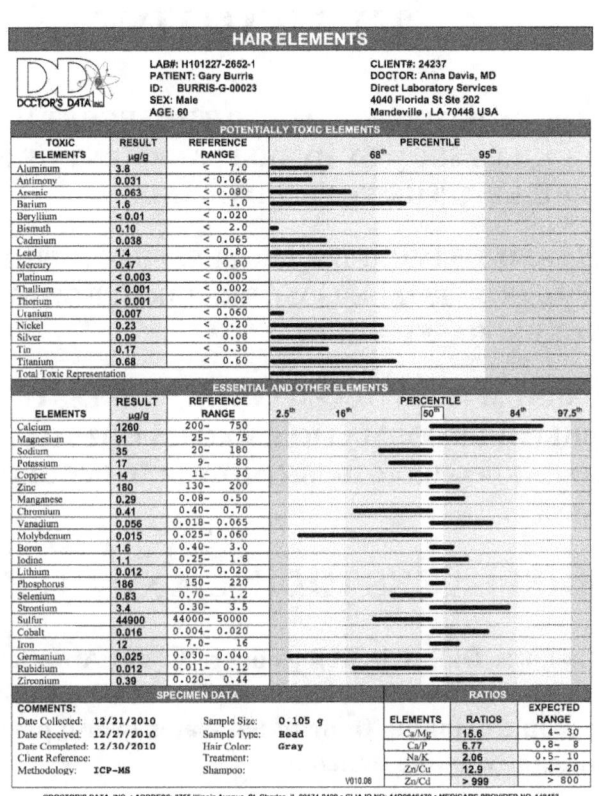

HAIR ELEMENTS

DOCTOR'S DATA INC

LAB#: H101227-2652-1
PATIENT: Gary Burris
ID: BURRIS-G-00023
SEX: Male
AGE: 60

CLIENT#: 24237
DOCTOR: Anna Davis, MD
Direct Laboratory Services
4040 Florida St Ste 202
Mandeville , LA 70448 USA

POTENTIALLY TOXIC ELEMENTS

TOXIC ELEMENTS	RESULT µg/g	REFERENCE RANGE	PERCENTILE 68th 95th
Aluminum	3.8	< 7.0	
Antimony	0.031	< 0.066	
Arsenic	0.063	< 0.080	
Barium	1.6	< 1.0	
Beryllium	< 0.01	< 0.020	
Bismuth	0.10	< 2.0	
Cadmium	0.038	< 0.065	
Lead	1.4	< 0.80	
Mercury	0.47	< 0.80	
Platinum	< 0.003	< 0.005	
Thallium	< 0.001	< 0.002	
Thorium	< 0.001	< 0.002	
Uranium	0.007	< 0.060	
Nickel	0.23	< 0.20	
Silver	0.09	< 0.08	
Tin	0.17	< 0.30	
Titanium	0.68	< 0.60	
Total Toxic Representation			

ESSENTIAL AND OTHER ELEMENTS

ELEMENTS	RESULT µg/g	REFERENCE RANGE	PERCENTILE 2.5th 16th 50th 84th 97.5th
Calcium	1260	200– 750	
Magnesium	81	25– 75	
Sodium	35	20– 180	
Potassium	17	9– 80	
Copper	14	11– 30	
Zinc	180	130– 200	
Manganese	0.29	0.08– 0.50	
Chromium	0.41	0.40– 0.70	
Vanadium	0.056	0.018– 0.065	
Molybdenum	0.015	0.025– 0.060	
Boron	1.6	0.40– 3.0	
Iodine	1.1	0.25– 1.8	
Lithium	0.012	0.007– 0.020	
Phosphorus	186	150– 220	
Selenium	0.83	0.70– 1.2	
Strontium	3.4	0.30– 3.5	
Sulfur	44900	44000– 50000	
Cobalt	0.016	0.004– 0.020	
Iron	12	7.0– 16	
Germanium	0.025	0.030– 0.040	
Rubidium	0.012	0.011– 0.12	
Zirconium	0.39	0.020– 0.44	

SPECIMEN DATA

COMMENTS:
Date Collected: 12/21/2010 — Sample Size: 0.105 g
Date Received: 12/27/2010 — Sample Type: Head
Date Completed: 12/30/2010 — Hair Color: Gray
Client Reference: — Treatment:
Methodology: ICP-MS — Shampoo:

V010.06

RATIOS

ELEMENTS	RATIOS	EXPECTED RANGE
Ca/Mg	15.6	4– 30
Ca/P	6.77	0.8– 8
Na/K	2.06	0.5– 10
Zn/Cu	12.9	4– 20
Zn/Cd	> 999	> 800

My hair sample showed me a very important thing.

I was low on sulfur.

Sulfur is the activator of our immune system.

Without sulfur, bad imbalance can happen inside the body.

FLUORIDE

FLUORIDE HAS BEEN CALLED THE SECRET WEAPON THAT KILLS SLOWLY BY DISPLACEMENT OF IODINE AND KILLING THE THYROID GLAND.

IT IS USED TO DOCILIZE PEOPLE. THE METHOD USED BY THE CONTROLLERS WORKS LIKE PROZAC, A FLUORIDE BASED DRUG.

ALUMINUM IS INVOLVED IN THE PROCESS OF ALZHEIMERS FORMATION. EXCESS CO2 THAT TURNS INTO CARBONIC ACID IS THE GLUE THAT CLOGS OUR BRAINS

DISPLACES IODINE AND LITHIUM.

Just a note: Dallas as well as over 70% of American cities' water is fluoridated. Fluoride replaces iodine in the thyroid and people drinking fluoridated water over a period will be significantly iodine deficient.

Fluoride also accumulates in the bones causing brittleness and eventually fused vertebrae and crippling skeletal fluorosis. The situation is exasperated by the diminished sunlight and lack of beneficial vitamin D.

Meanwhile harmful UV radiation is increasing. As reported by the *Weston Price Foundation*, we need Vitamin K to metabolize Vitamin D. Vitamin K is in foods such as grass-fed butter, fermented cod liver

oil and is **severely deficient in our diets.**

Fluoride, Teeth, and the Atomic Bomb

by Chris Bryson & Joel Griffiths

Introduction: The following article was commissioned by the Christian Science Monitor in the spring of 1997. Despite much favorable comment from editors, and full

documentation, the story remains unpublished by the Monitor. By any yardstick, this report was an award-winning scoop for any national paper. The report offers a glimpse into the history of fluoride, a bio-accumulative toxic that Americans ingest every day.

The authors, Griffiths and Bryson, spent more than a year on research. With the belief that the information should be withheld no longer, the authors gave their report to Waste Not, and others, with a short note: "use as you wish."

This introduction is taken from Waste Not #414 (September 1997) where the article was first published. The article went on to be nominated as the year's 18th most censored story in the 1998 Project Censored Series.

Fluoride, Teeth, and the Atomic Bomb
by Chris Bryson & Joel Griffiths

Some fifty years after the United States began adding fluoride to public water supplies to reduce cavities in children's teeth, declassified government documents are shedding new light on the roots of that still-controversial public health measure, revealing a surprising connection between fluoride and the dawning of the nuclear age.

Today, two thirds of U.S. public drinking water supplies are fluoridated. Many municipalities still resist the practice, disbelieving the government's assurances of safety.

Since the days of World War II, when this nation prevailed

by building the world's first atomic bomb, U.S. public health leaders have maintained that low doses of fluoride are safe for people, and good for children's teeth.

That safety verdict should now be re-examined in the light of hundreds of once-secret WWII documents obtained by Griffiths and Bryson –including declassified papers of the

Manhattan Project, the U.S. military group that built the atomic bomb.

Fluoride was the key chemical in atomic bomb production, according to the documents. Massive quantities of fluoride– millions of tons– were essential for the manufacture of bomb-grade uranium and plutonium for nuclear weapons throughout the Cold War. One of the most toxic chemicals known, fluoride rapidly emerged as the leading chemical health hazard of the U.S atomic bomb program–both for workers and for nearby communities, the documents reveal.

Other revelations include:

*** Much of the original proof that fluoride is safe for humans in low doses was generated by A-bomb program scientists, who had been secretly ordered to provide "evidence useful in litigation" against defense contractors for fluoride injury to citizens. The first lawsuits against the U.S. A-bomb program were not over radiation, but over fluoride damage,**

*** Human studies were required. Bomb program researchers played a leading role in the design and implementation of the most extensive U.S. study of the health effects of**

fluoridating public drinking water–conducted in Newburgh, New York from 1945 to 1956. Then, in a classified operation code-named "Program F," they secretly gathered and analyzed blood and tissue samples from Newburgh citizens, with the cooperation of State Health Department personnel.

*** The original secret version–obtained by these reporters–of a 1948 study published by Program F scientists in the**

Journal of the American Dental Association shows that evidence of adverse health effects from fluoride was censored by the U.S. Atomic Energy Commission (AEC) – considered the most powerful of Cold War agencies– for reasons of national security.

*** The bomb program's fluoride safety studies were conducted at the University of Rochester, site of one of the most notorious human radiation experiments of the Cold War, in which unsuspecting hospital patients were injected with toxic doses of radioactive plutonium. The fluoride studies were conducted with the same ethical mind-set, in which "national security" was paramount.**

*** The U.S. government's conflict of interest–and its motive to prove fluoride "safe" — has not until now been made clear to the public in the furious debate over water fluoridation since the 1950′s, nor to civilian researchers and health professionals, or journalists.**

The declassified documents resonate with a growing body of scientific evidence, and a chorus of questions, about the

health effects of fluoride in the environment.

Human exposure to fluoride has mushroomed since World War II, due not only to fluoridated water and toothpaste, but to environmental pollution by major industries from aluminum to pesticides: fluoride is a critical industrial chemical.

The impact can be seen, literally, in the smiles of our children. Large numbers of U.S. young people–up to 80 percent in some cities–now have dental fluorosis, the first visible sign of excessive fluoride exposure, per the U.S. National Research Council. (The signs are whitish flecks or spots, particularly on the front teeth, or dark spots or stripes in more severe cases.)

Less-known to the public is that fluoride also accumulates in bones –"The teeth are windows to what's happening in the bones," explains Paul Connett, Professor of Chemistry at St. Lawrence University (N.Y.). In recent years, pediatric bone specialists have expressed alarm about an increase in stress fractures among U.S. young people. Connett and other scientists are concerned that fluoride –linked to bone damage by studies since the 1930′s– may be a contributing factor. The declassified documents add urgency: much of the original proof that low-dose fluoride is safe for children's bones came from U.S. bomb program scientists, according to this investigation.

Now, researchers who have reviewed these declassified documents fear that Cold War national security

considerations may have prevented objective scientific evaluation of vital public health questions concerning fluoride.

"Information was buried," concludes Dr. Phyllis Mullenix, former head of toxicology at Forsyth Dental Center in Boston, and now a critic of fluoridation. Animal studies Mullenix and co-workers conducted at Forsyth in the early 1990's indicated that fluoride was a powerful central nervous system (CNS) toxin, and might adversely affect human brain functioning, even at low doses.

(New epidemiological evidence from China adds support, showing a correlation between low-dose fluoride exposure and diminished I.Q. in children.) Mullenix's results were published in 1995, in a reputable peer-reviewed scientific journal.

During her investigation, Mullenix was astonished to discover there had been virtually no previous U.S. studies of fluoride's effects on the human brain. Then, her application for a grant to continue her CNS research was turned down by the U.S. National Institutes of Health (NIH), where an NIH panel, she says, flatly told her that "fluoride does not have central nervous system effects."

Declassified documents of the U.S. atomic-bomb program indicate otherwise. An April 29, 1944 Manhattan Project memo reports: "Clinical evidence suggests that uranium hexafluoride may have a rather marked central nervous system effect…. It seems most likely that the F [code for

fluoride] component rather than the T [code for uranium] is the causative factor."

The memo --stamped "secret"-- is addressed to the head of the Manhattan Project's Medical Section, Colonel Stafford Warren. Colonel Warren is asked to approve a program of animal research on CNS effects: "Since work with these compounds is essential, it will be necessary to know in advance what mental effects may occur after exposure...This is important not only to protect a given individual, but also to prevent a confused workman from injuring others by improperly performing his duties."

On the same day, Colonel Warren approved the CNS research program. This was in 1944, at the height of the Second World War and the nation's race to build the world's first atomic bomb. For research on fluoride's CNS effects to be approved at such a momentous time, the supporting evidence set forth in the proposal forwarded along with the memo must have been persuasive.

The proposal, however, is missing from the files of the U.S. National Archives. "If you find the memos, but the document they refer to is missing, it's probably still classified," said Charles Reeves, chief librarian at the Atlanta branch of the U.S. National Archives and Records Administration, where the memos were found. Similarly, no results of the Manhattan Project's fluoride CNS research could be found in the files.

After reviewing the memos, Mullenix declared herself "flabbergasted." She went on, "How could I be told by NIH that fluoride has no central nervous system effects when these documents were sitting there all the time?" She reasons that the Manhattan Project did do fluoride CNS studies --"that kind of warning, that fluoride workers might be a danger to the bomb program by improperly performing their duties--I can't imagine that would be ignored"-- but that the results were buried because they might create a difficult legal and public relations problem for the government.

The author of the 1944 CNS research proposal was Dr. Harold C. Hodge, at the time chief of fluoride toxicology studies for the University of Rochester division of the Manhattan Project. Nearly fifty years later at the Forsyth Dental Center in Boston, Dr. Mullenix was introduced to a gently ambling elderly man brought in to serve as a consultant on her CNS research--Harold C. Hodge. By then Hodge had achieved status emeritus as a world authority on fluoride safety. "But even though he was supposed to be helping me," says Mullenix, "he never once mentioned the CNS work he had done for the Manhattan Project."

The "black hole" in fluoride CNS research since the days of the Manhattan Project is unacceptable to Mullenix, who refuses to abandon the issue. "There is so much fluoride exposure now, and we simply do not know what it is doing," she says. "You can't just walk away from this."

Dr. Antonio Noronha, an NIH scientific review advisor familiar with Dr. Mullenix's grant request, says her proposal was rejected by a scientific peer-review group. He terms her claim of institutional bias against fluoride CNS research "farfetched." He adds, "We strive very hard at NIH to make sure politics does not enter the picture."

Fluoride and National Security

The documentary trail begins at the height of WW2, in 1944, when a severe pollution incident occurred downwind of the E.I. du Pont du Nemours Company chemical factory in Deepwater, New Jersey. The factory was then producing millions of pounds of fluoride for the Manhattan project, the ultra-secret U.S. military program racing to produce the world's first atomic bomb.

The farms downwind in Gloucester and Salem counties were famous for their high-quality produce -- their peaches went directly to the Waldorf Astoria Hotel in New York. Their tomatoes were bought up by Campbell's Soup.

But in the summer of 1943, the farmers began to report that their crops were blighted, and that "something is burning up the peach crops around here."

Poultry died after an all-night thunderstorm, they reported. Farm workers who ate the produce they had picked sometimes vomited all night and into the next day. "I remember our horses looked sick and were too stiff to work," these reporters were told by Mildred Giordano, who

was a teenager at the time. Some cows were so crippled they could not stand up, and grazed by crawling on their bellies.

The account was confirmed in taped interviews, shortly before he died, with Philip Sadtler of Sadtler Laboratories of Philadelphia, one of the nation's oldest chemical consulting firms. Sadtler had personally conducted the initial investigation of the damage.

Although the farmers did not know it, the attention of the Manhattan Project and the federal government was riveted on the New Jersey incident, according to once-secret documents obtained by these reporters. After the war's end, in a secret Manhattan Project memo dated March 1, 1946, the Project's chief of fluoride toxicology studies, Harold C. Hodge, worriedly wrote to his boss Colonel Stafford L. Warren, Chief of the Medical Division, about "problems associated with the question of fluoride contamination of the atmosphere in a certain section of New Jersey. There seem to be four distinct (though related) problems," continued Hodge;

1. A question of injury of the peach crop in 1944.

2. A report of extraordinary fluoride content of vegetables grown in this area.

3. A report of abnormally high fluoride content in the blood of human individuals residing in this area.

4. A report raising the question of serious poisoning of

horses and cattle in this area.

The New Jersey farmers waited until the war was over, then sued du Pont and the Manhattan Project for fluoride damage -- reportedly the first lawsuits against the U.S. A-bomb program.

Although seemingly trivial, the lawsuits shook the government, the secret documents reveal. Under the personal direction of Manhattan Project chief Major General Leslie R.Groves, secret meetings were convened in Washington, with compulsory attendance by scores of scientists and officials from the U.S War Department, the Manhattan Project, the Food and Drug Administration, the Agriculture and Justice Departments, the U.S Army's Chemical Warfare Service and Edgewood Arsenal, the Bureau of Standards, and du Pont lawyers. Declassified memos of the meetings reveal a secret mobilization of the full forces of the government to defeat the New Jersey farmers:

These agencies "are making scientific investigations to obtain evidence which may be used to protect the interest of the Government at the trial of the suits brought by owners of peach orchards in ... New Jersey," stated Manhattan Project Lieutenant Colonel Cooper B. Rhodes, in a memo c.c.'d to General Groves.

27 August 1945

Subject: Investigation of Crop Damage at Lower Penns Neck, New Jersey
To: The Commanding General, Army Service Forces, Pentagon Building, Washington D.C.

"At the request of the Secretary of War the Department of Agriculture has agreed to cooperate in investigating complaints of crop damage attributed... to fumes from a plant operated in connection with the Manhattan Project."

Signed, L.R. Groves, Major General U.S.

"The Department of Justice is cooperating in the defense of these suits," wrote General Groves in a Feb. 28, 1946 memo to the Chairman of the U.S. Senate Special Committee on Atomic Energy.

Why the national-security emergency over a few lawsuits by New Jersey farmers? In 1946 the United States had begun full-scale production of atomic bombs. No other nation had yet tested a nuclear weapon, and the A-bomb was seen as crucial for U.S leadership of the postwar world. The New Jersey fluoride lawsuits were a serious roadblock to that strategy.

"The specter of endless lawsuits haunted the military," writes Lansing Lamont in his acclaimed book about the first atomic bomb test, "Day of Trinity."

In the case of fluoride, "If the farmers won, it would open the door to further suits, which might impede the bomb

program's ability to use fluoride," said Jacqueline Kittrell, a Tennessee public interest lawyer specializing in nuclear cases, who examined the declassified fluoride documents. (Kittrell has represented plaintiffs in several human radiation experiment cases.) She added, "The reports of human injury were especially threatening, because of the potential for enormous settlements -- not to mention the PR problem."

Indeed, du Pont was particularly concerned about the "possible psychologic reaction" to the New Jersey pollution incident, according to a secret 1946 Manhattan Project memo. Facing a threat from the Food and Drug Administration (FDA) to embargo the region's produce because of "high fluoride content," du Pont dispatched its lawyers to the FDA offices in Washington, where an agitated meeting ensued. According to a memo sent next day to

General Groves, Du Pont's lawyer argued "that in view of the pending suits...any action by the Food and Drug Administration... would have a serious effect on the du Pont Company and would create a bad public relations situation." After the meeting adjourned, Manhattan Project Captain John Davies approached the FDA's Food Division chief and "impressed upon Dr. White the substantial interest which the Government had in claims which might arise as a result of action which might be taken by the Food and Drug Administration."

There was no embargo. Instead, new tests for fluoride in the New Jersey area would be conducted -- not by the

Department of Agriculture -- but by the U.S. Army's Chemical Warfare Service because "work done by the Chemical Warfare Service would carry the greatest weight as evidence if... lawsuits are started by the complainants." The memo was signed by General Groves.

Meanwhile, the public relations problem remained unresolved -- local citizens were in a panic about fluoride.

The farmer's spokesman, Willard B. Kille, was personally invited to dine with General Groves --then known as "the man who built the atomic bomb" -- at his office at the War Department on March 26, 1946. Although he had been diagnosed with fluoride poisoning by his doctor, Kille departed the luncheon convinced of the government's good faith. The next day he wrote to the general, wishing the other farmers could have been present, he said, so "they too could come away with the feeling that their interests in this particular matter were being safeguarded by men of the very highest type whose integrity they could not question."

In a subsequent secret Manhattan project memo, a broader solution to the public relations problem was suggested by chief fluoride toxicologist Harold C. Hodge. He wrote to the Medical Section chief, Col. Warren: "Would there be any use in making attempts to counteract the local fear of fluoride on the part of residents of Salem and Gloucester counties through lectures on F toxicology and perhaps the usefulness of F in tooth health?" Such lectures were indeed given, not only to New Jersey citizens but to the rest of the nation throughout the Cold War.

The New Jersey farmers' lawsuits were ultimately stymied by the government's refusal to reveal the key piece of information that would have settled the case --how much fluoride du Pont had vented into the atmosphere during the war. "Disclosure... would be injurious to the military security of the United States," wrote Manhattan Project Major C.A Taney, Jr. The farmers were pacified with token financial settlements, according to interviews with descendants still living in the area.

"All we knew is that du Pont released some chemical that burned up all the peach trees around here," recalls Angelo Giordano, whose father James was one of the original plaintiffs. "The trees were no good after that, so we had to give up on the peaches." Their horses and cows, too, acted stiff and walked stiff, recalls his sister Mildred. "Could any of that have been the fluoride?" she asked. (The symptoms she detailed to the authors are cardinal signs of fluoride toxicity, according to veterinary toxicologists.)

The Giordano family, too, has been plagued by bone and joint problems, Mildred adds. Recalling the settlement received by the Giordanos, Angelo told these reporters that "my father said he got about $200."

The farmers were stonewalled in their search for information, and their complaints have long since been forgotten. But they unknowingly left their imprint on history -- their claims of injury to their health reverberated through the corridors of power in Washington, and

triggered intensive secret bomb-program research on the health effects of fluoride. A secret 1945 memo from Manhattan Project Lt. Col. Rhodes to General Groves stated: "Because of complaints that animals and humans have been injured by hydrogen fluoride fumes in [the New Jersey] area, although there are no pending suits involving such claims, the University of Rochester is conducting experiments to determine the toxic effect of fluoride."

Much of the proof of fluoride's safety in low doses rests on the postwar work performed by the University of Rochester, in anticipation of lawsuits against the bomb program for human injury.

Fluoride and the Cold War.

Delegating fluoride safety studies to the University of Rochester was not surprising. During WWII the federal government had become involved, for the first time, in large-scale funding of scientific research at government-owned labs and private colleges. Those early spending priorities were shaped by the nation's often-secret military needs.

The prestigious upstate New York college had housed a key wartime division of the Manhattan Project, studying the health effects of the new "special materials," such as uranium, plutonium, beryllium and fluoride, being used to make the atomic bomb. That work continued after the war, with millions of dollars flowing from the Manhattan Project and its successor organization, the Atomic Energy

Commission (AEC). (Indeed, the bomb left an indelible imprint on all U.S. science in the late 1940's and 50's. Up to 90% of federal funds for university research came from either the Defense Department or the AEC in this period, according to Noam Chomsky's 1996 book "The Cold War and the University.")

The University of Rochester medical school became a revolving door for senior bomb program scientists. Postwar faculty included Stafford Warren, the top medical officer of the Manhattan Project, and Harold Hodge, chief of fluoride research for the bomb program.

But this marriage of military secrecy and medical science bore deformed offspring. The University of Rochester's classified fluoride studies -- code- named Program F -- were conducted at its Atomic Energy Project (AEP), a top-secret facility funded by the AEC and housed in Strong Memorial Hospital. It was there that one of the most notorious human radiation experiments of the Cold War took place, in which unsuspecting hospital patients were injected with toxic doses of radioactive plutonium. Revelation of this experiment in a Pulitzer prize-winning account by Eileen Welsome led to a 1995 U.S. Presidential investigation, and a multimillion-dollar cash settlement for victims.

Program F was not about children's teeth. It grew directly out of litigation against the bomb program and its main purpose was to furnish scientific ammunition which the government and its nuclear contractors could use to defeat lawsuits for human injury. Program F's director was none other than Harold C. Hodge, who had led the Manhattan Project investigation of alleged human injury in the New Jersey fluoride-pollution incident.

Program F's purpose is spelled out in a classified 1948 report. It reads: "To supply evidence useful in the litigation arising from an alleged loss of a fruit crop several years ago, a number of problems have been opened. Since excessive blood fluoride levels were reported in human residents of the same area, our principal effort has been devoted to describing the relationship of blood fluorides to toxic effects."

The litigation referred to, of course, and the claims of human injury were against the bomb program and its contractors. Thus, the purpose of Program F was to obtain evidence useful in litigation against the bomb program. The research was being conducted by the defendants.

The potential conflict of interest is clear. If lower dose ranges were found hazardous by Program F, it might have opened the bomb program and its contractors to lawsuits for injury to human health, as well as public outcry.

Comments lawyer Kittrell: "This and other documents indicate that the University of Rochester's fluoride research grew out of the New Jersey lawsuits and was performed in anticipation of lawsuits against the bomb program for human injury. Studies undertaken for litigation purposes by the defendants would not be considered scientifically

acceptable today, " adds Kittrell, "because of their inherent bias to prove the chemical safe."

Unfortunately, much of the proof of fluoride's safety rests on the work performed by Program F Scientists at the University of Rochester. During the postwar period that university emerged as the leading academic center for establishing the safety of fluoride, as well as its effectiveness in reducing tooth decay, according to Dental School spokesperson William H. Bowen, MD. The key figure in this research, Bowen said, was Harold C. Hodge-- who also became a leading national proponent of fluoridating public drinking water. Program F's interest in water fluoridation was not just 'to counteract the local fear of fluoride on the part of residents,' as Hodge had earlier written. The bomb program needed human studies, as they had needed human studies for plutonium, and adding fluoride to public water supplies provided one opportunity.

The A-Bomb Program and Water Fluoridation

Bomb-program scientists played a prominent -- if unpublicized -- role in the nation's first-planned water fluoridation experiment, in Newburgh, New York. The Newburgh Demonstration Project is considered the most extensive study of the health effects of fluoridation, supplying much of the evidence that low doses are safe for children's bones, and good for their teeth.

Planning began in 1943 with the appointment of a special New York State Health Department committee to study the

advisability of adding fluoride to Newburgh's drinking water. The chairman of the committee was Dr. Hodge, then chief of fluoride toxicity studies for the Manhattan Project.

Subsequent members included Henry L. Barnett, a captain in the Project's Medical section, and John W. Fertig, in 1944 with the office of Scientific Research and Development, the Pentagon group which sired the Manhattan Project. Their military affiliations were kept secret: Hodge was described as a pharmacologist, Barnett as a pediatrician. Placed in charge of the Newburgh project was David B. Ast, chief dental officer of the State Health Department. Ast had participated in a key secret wartime conference **on fluoride held by the Manhattan Project, and later worked with Dr. Hodge on the Project's investigation of human injury in the New Jersey incident, according to once-secret memos.**

The committee recommended that Newburgh be fluoridated. It also selected the types of medical studies to be done, and "provided expert guidance" for the duration of the experiment. The key question to be answered was: "Are there any cumulative effects -- beneficial or otherwise, on tissues and organs other than the teeth -- of long-continued ingestion of such small concentrations...?"

According to the declassified documents, this was also key information sought by the bomb program, which would require long-continued exposure of workers and communities to fluoride throughout the Cold War.

In May 1945, Newburgh's water was fluoridated, and over

the next ten years its residents were studied by the State Health Department. In tandem, Program F conducted its own secret studies, focusing on the amounts of fluoride Newburgh citizens retained in their blood and tissues - key information sought by the bomb program: "Possible toxic effects of fluoride were in the forefront of consideration," the advisory committee stated. Health Department personnel cooperated, shipping blood and placenta samples to the Program F team at the University of Rochester. The samples were collected by Dr. David B. Overton, the Department's chief of pediatric studies at Newburgh.

The final report of the Newburgh Demonstration Project, published in 1956 in the Journal of the American Dental Association, concluded that "small concentrations" of fluoride were safe for U.S.citizens. The biological proof -- "based on work performed ... at the University of Rochester Atomic Energy Project" -- was delivered by Dr. Hodge.

Today, news that scientists from the atomic bomb program secretly shaped and guided the Newburgh fluoridation experiment, and studied the citizen's blood and tissue samples, is greeted with incredulity.

"I'm shocked -- beyond words," said present-day Newburgh Mayor Audrey Carey, commenting on these reporters' findings. "It reminds me of the Tuskegee experiment that was done on syphilis patients down in Alabama."

As a child in the early 1950's, Mayor Carey was taken to the old firehouse on Broadway in Newburgh, which housed the

Public Health Clinic. There, doctors from the Newburgh fluoridation project studied her teeth, and a peculiar fusion of two finger bones on her left hand she had been born with. Today, adds Carey, her granddaughter has white dental-fluorosis marks on her front teeth.

Mayor Carey wants answers from the government about the secret history of fluoride, and the Newburgh fluoridation experiment. "I absolutely want to pursue it," she said. "It is appalling to do any kind of experimentation and study without people's knowledge and permission."

Contacted by these reporters, the director of the Newburgh experiment, David B. Ast, says he was unaware Manhattan Project scientists were involved. "If I had known, I would have been certainly investigating why, and what the connection was," he said. Did he know that blood and placenta samples from Newburgh were being sent to bomb program researchers at the University of Rochester? "I was not aware of it," Ast replied. Did he recall participating in the Manhattan Project's secret wartime conference on fluoride in January 1944, or going to New Jersey with Dr.

Hodge to investigate human injury in the du Pont case--as secret memos state? He told the reporters he had no recollection of these events.

A spokesperson for the University of Rochester Medical Center, Bob Loeb, confirmed that blood and tissue samples from Newburgh had been tested by the University's Dr. Hodge. On the ethics of secretly studying U.S citizens to obtain information useful in litigation against the A-bomb program, he said, "that's a question we cannot answer." He referred inquiries to the U.S. Department of Energy (DOE), successor to the Atomic Energy Commission.

A spokesperson for the DOE in Washington, Jayne Brady, confirmed that a review of DOE files indicated that a "significant reason" for fluoride experiments conducted at the University of Rochester after the war was "impending litigation between the du Pont company and residents of New Jersey areas." However, she added, "DOE has found no documents to indicate that fluoride research was done to protect the Manhattan Project or its contractors from lawsuits."

On Manhattan Project involvement in Newburgh, the spokesperson stated, "Nothing that we have suggests that the DOE or predecessor agencies -- especially the Manhattan Project -- authorized fluoride experiments to be performed on children in the 1940's."

When told that the reporters had several documents that directly tied the Manhattan Project's successor agency at the University of Rochester, the AEP, to the Newburgh experiment, the DOE spokesperson later conceded her study was confined to "the available universe" of documents. Two days later spokesperson Jayne Brady faxed a statement for clarification: "My search only involved the documents that we collected as part of our human radiation experiments project -- fluoride was not part of our research effort.

"Most significantly," the statement continued, relevant documents may be in a classified collection at the DOE Oak Ridge National Laboratory known as the Records Holding Task Group. "This collection consists entirely of classified documents removed from other files for the purpose of classified document accountability many years ago," and was "a rich source of documents for the human radiation experiments project," she said.

The crucial question arising from this investigation is: Were adverse health findings from Newburgh and other bomb-program fluoride studies suppressed? All AEC-funded studies had to be declassified before publication in civilian medical and dental journals. Where are the original classified versions?

The transcript of one of the major secret scientific conferences of WW2--on "fluoride metabolism"--is missing from the files of the U.S. National Archives. Participants in the conference included key figures who promoted the safety of fluoride and water fluoridation to the public after the war - Harold Hodge of the Manhattan Project, David B. Ast of the Newburgh Project, and U.S. Public Health Service

Dentist H.Trendley Dean, popularly known as the "father of fluoridation." "If it is missing from the files, it is probably still classified," National Archives librarians told these reporters.

A 1944 WW2 Manhattan Project classified report on water fluoridation is missing from the files of the University of

Rochester Atomic Energy Project, the U.S. National Archives, and the Nuclear Repository at the University of Tennessee, Knoxville. The next four numerically consecutive documents are also missing, while the remainder of the "MP-1500 series" is present. "Either those documents are still classified, or they've been 'disappeared' by the government," says Clifford Honicker, Executive Director of the American Environmental Health Studies Project, in Knoxville, Tennessee, which provided key evidence in the public exposure and prosecution of U.S. human radiation experiments.

Seven pages have been cut out of a 1947 Rochester bomb-project notebook entitled "Du Pont litigation." "Most unusual," commented chief medical school archivist Chris Hoolihan.

Similarly, Freedom of Information Act (FOIA) requests by these authors over a year ago with the DOE for hundreds of classified fluoride reports have failed to dislodge any. "We're behind," explained Amy Rothrock, FOIA officer for the Department of Energy at their Oak Ridge operations.

Was information suppressed? These reporters made what appears to be the first discovery of the original classified version of a fluoride safety study by bomb program scientists.

A censored version of this study was later published in the August 1948 Journal of the American Dental Association. Comparison of the secret with the published version

indicates that the U.S. AEC did censor damaging information on fluoride, to the point of tragicomedy.

This was a study of the dental and physical health of workers in a factory producing fluoride for the A-bomb program, conducted by a team of dentists from the Manhattan Project.

*** The secret version reports that most of the men had no teeth left. The published version reports only that the men had fewer cavities.**

*** The secret version says the men had to wear rubber boots because the fluoride fumes disintegrated the nails in their shoes. The published version does not mention this.**

*** The secret version says the fluoride may have acted similarly on the men's teeth, contributing to their 'toothlessness'. The published version omits this statement.**

The published version concludes that "the men were unusually healthy, judged from both a medical and dental point of view."

Asked for comment on the early links of the Manhattan

Project to water fluoridation, Dr Harold Slavkin, Director of the National Institute for Dental Research, the U.S. agency which today funds fluoride research, said, "I wasn't aware of any input from the Atomic Energy Commission." Nevertheless, he insisted, fluoride's efficacy and safety in the

prevention of dental cavities over the last fifty years is well-proved. "The motivation of a scientist is often different from the outcome, " he reflected. "I do not hold a prejudice about where the knowledge comes from."

After comparing the secret and published versions of the censored study, toxicologist Phyllis Mullenix commented, "This makes me ashamed to be a scientist." Of other Cold War-era fluoride safety studies, she asks, "Were they all done like this?"

Doctor Martine Carroll PHD

A Trained Phycologist looks at the evidence of Chemtrails

During the process of acceptance of bad news, we go through physical and emotional shocks in stages with lessening emotional trauma

Denial, anger, depression, grief then acceptance are the stages we go through all of us as we try to wrap our minds around these issues.

Question to Martine

You are a physiologist! And you see actions that destroy the life support system and they continue to do, Is that sane or insane?

Martine Carroll

"Well it is pathological and it is insane!

What is amazing is that the information cannot be fathomed by the people reading it they cannot understand it even though it is happening and is in understandable terms.

They still do not understand IT.

MACONDO DOWNWINDERS

British Petro and the toxic killing go on

The Macondo disaster changed everything for man and beast stunning evidence, in real time, has shown that there is a health disaster time bomb ticking along the coast and far inland. One of many symptoms of chemically induced mental illness is paralysis of emotion and instincts.

After long detoxification and mineral replacement, my story of BP exposures is the story of the people over a very wide area locally and far downwind.

My health issues from repeated close exposure to high levels of their poison cost me a year of my life in decline and then another in recovery from something dark after I discovered what had been done to us, The Macondo Down winders. Many are far more ill and some have died: This is their story. After I became very ill repeatedly I wanted to know what had been done to me first, then by extension the coastal community and the food providing environment in general.

Long and dense exposures to dispersant spray and oil smoke had rendered personal eco-systems or immune systems, or people and other mammals, ready for any disease. The toxins affected me, and many others for months at a time with fatigue, sever lung impairment, and numerous toxic-related symptoms and affected many others, ability to work for long periods of time and the BP event limited the area of life sustaining estuaries and the amount of harvesting work due to extreme toxins in the air and water for months: And the perception that food contamination issues were being understated by authorities.

Today the sunk oil and the dispersant components

surface every time the tide is low and the water warms or a storm stirs up the evidence and another silent chemical attack invade the bodies of all living downwind.

Downwind includes most of middle and eastern America and Canada eventually reaching Europe then circling back to the source in Louisiana. And as of May 2012, the spraying still goes on under the cover of dark, but not completely out of the eye of the beholder. The spraying continues because of under reported leaks from the well and 27,000 other old wells now leaking and more cracks in the sea bottom.

One could read one news report saying 27,000 natural leaks and another more informed media source would state 27,000 old wells unmaintained and leaking. Everything you heard during the media circus was suspect.

Given my decades of information gathering and filmmaking and commercial fishing around the gulf, and the ocean world, I had seen the old leaks up-close. The oil industry had always been a nasty partner in our collective energy gathering habits.

During the recent season opening in June 2012, Of the few boats still crossing the now sickened St. Bernard waters, the 'Lil Ric' and its seasoned crew saw that things were not the way the story is still being told.

Rick called from the boat in May 2010 and said "whoosh" was what he heard when they flew near us anchored behind the out islands, whoosh and then the smell." Whoosh and the secrets go on as corporate treason and the takeover of the US government was getting clearer to more people.

With each passing month, with each new obit with the name of the young to middle aged in majority: Everyone knew someone who died young in St. Bernard since Macondo. The spiritual deaths and injuries of Katrina made the Macondo madness seem like the preverbal "straw" that broke the camel's back. These, some unnaturally strong people, of the parish have my respects and gratitude.

Now, June 2012, everyone has some health problem on the bayou and the story up town is the same. Strong fishermen who would "work three days around the clock on deck and felt good, now a short

walk to the dock carrying groceries for a trip "tires me" and it happened quick after we got hit with heavy poison and smoke."

This became a common comment everywhere along the coast and among the fishermen and our natural food providers.

In New Orleans, I was told by a confidant in state government that a group of middle aged women had all developed anal bleeding after weeks, by June, of heavy chemical smell over uptown that was coming on the wind from the south from the spraying of Macondo oil.

I met and interviewed one of these "anal bleeders". She told me many of her friends now take medications, from time to time, for the conditions caused by exposures to extreme levels of neurotoxins and other cell and nerve damaging agents. She also insisted that the low level dispersant planes were flying at low, low altitude directly over the city at night.

This was an act of cowards' war: Since the Katrina exodus New Orleans, and the oil coast, has been repopulated by more than 35% Germanic subtle invaders.

Yes, the invasion is that real. Invade with a weather war, destroy, get great funds to rebuilt infrastructure. Pay no one for health effects and keep this reality out of court. I was the first reported, across world media, poisoning from "Corexist".

Three law firms I hired were BP retained (secret weapon) lawyers and they all stalled me and screwed me and made millions keeping my evidence in check. Obviously, this tactic did not work I am stilling standing and writing and making the next mega documentary.

I saw this over low level spraying over New Orleans. The HAAD program was evident over the city for many years.

Her husband worked for an oil company and she was unwilling to go on camera and refused to let me use here name in the story. When I shared with her, a many generation Jew, the information that German invaders were hiding as Jews and were heavily involved in the HAAD or high altitude aerosol dispersant program.

They had been involved for decades. She immediately banned all reference to "chemo-trails" on her large face book information exchange site that she used to organize protests in the area. Some schools in Louisiana, now keep inhalers at school due to the many lung problems with children. No study concerning the long-term effects of the spraying has ever been released to the public.

If the obituaries are correct our children will die younger in this world depopulation scheme. It is now playing out along the entire Gulf coast.

If Jesse Ventura got the story right this may be the last semi-healthy generation to fish the bayous or even live there.

I recall the interview comment from the Dallas doctor telling me "children of this generation will not outlive their parents".

It is like these depopulation people, some now working in the White House, half admit to this shit when feeding the media bits and pieces of information.

They seem to know most of us can no longer react to the danger. It is like the fight or flight response

was turned off along the neurotoxin-ed coastal and uplands areas.

From my direct observation, much of the gentile world and rank and file Jewish people have that reality switch turned off chemically.

I recall Ricky's Robins' story of the pelicans on the Chandelier Barrier islands not trying to get away within days of the spill and irrational spraying:

These creatures just accepted their fate without question and as the poison did its work on the minds of those near and far the people acted the same way. Too poisoned to know they are being poisoned.

Heavy sleep inducing exposures occurred to many in south St. Bernard Parish, Louisiana, and in the Mississippi Sound, where I was filming during the SPILL and I had worked there periodically since Katrina. EPA data many data surveys showed people were exposed all around the Gulf. Friends of mine in Homosassa springs reported that many retirement community people were dying suddenly. The toxic clouds went over all north Florida and eventually up the east coast and is still circling around Earth with the jet stream or it is in the

stratosphere, where there is no weather. Here it drifts down slowly for centuries.

While, on a shrimp boat with Captain Catfish, filming and exposing the sunk oil lie/story, me and my dogs were knocked out in minutes by the chemical cloud that surrounded us. This was off the Mississippi coast during one of my boat trips with him.

We were in the middle of a massive thin slick that the now obviously totally controlled government officials said was not there.

My dogs both got real sick and I got sicker as the weeks went by and my mind was so 'chargrilled' I did not desire to run away from it yet. One dog "caught cancer" when a growth came up within three days after a heavy spraying event and was saved with timely surgery.

The other got hit in the kidneys and both became incontinent on the same day after multiple spraying from high and low levels planes.

The man keeping them in Hopedale, Louisiana began vomiting blood at the same time and

eventually 'went insane'.

I had to struggle to not pee my pants many times. This passed after we all got on Arkansas spring water I got from my favorite spring without fluoride.

It took several weeks and an exodus back to Louisiana and then, more upwind, to Florida after finding Arkansas being sprayed by high level 'chemo' planes all summer and fall of 2010.

The HAAD spraying continues today all through the south and the whole country but that is another deeper story.

The next straw was seeing, with others, Air Force c-130 'dispersant' planes on Nov 8 2010 over North Little Rock, Arkansas spraying near and over the Airbase at Jacksonville.

Also, there were hundreds of chemtrails/chemo-trol flights all day, and all night, over Arkansas that turned the sky powder white.

Birds fell from the sky, fish died, children cried and horses bucked and went crazy after a heavy chemical snow visibly fell on Arkansas Sept.28,2010.

The forests were silent for days as the birds in the open environments headed somewhere away from this madness. Deer wandered unafraid in the yard of the cabin where I was staying in Arkansas while trying to escape the toxic cloud of Macondo. The cloud followed me like that Snuffy Smith character, again.

After the Arkansas bird kill incident, John Wheeler, former special Assistant secretary of the Air force under Bush, 2005-2008, knew something was happening with chemical weapons testing over populated areas in Arkansas and the bird's death were part of a test of an attack, or it was an attack. The same plane that "accidentally" did something that killed the birds.

The pilots had another "accident" after flying from Arkansas to Italy and released more poison and killed birds over that country.

The Former special assistant secretary went to the White house and the air force to complain.

The next day he was found dead in a dumpster in Delaware. In a very arcane way this story of HAAD circles about and flies through the Macondo smoke and mirrors.

Evidence points to the miscalculations of the effects of mixing the test poison with the gases present over the state for months until the first northern on Sept 28 2010 cleared the sky a little.

The test killed outright a great number of birds and fish and today man middle-aged men are dying from sudden heart failure.

This is consummate with the high levels of aluminum in the air that effectively poison the heart and the blood.

Many children brought to little Rocks children's medical center are dying quickly and as told to me by one nurse practitioner who quit "something is going on that is "making doctors rich. And they do want to ask why all the profitable death".

Little Rock seems to be a dying zone for our children already sicken by the fluoride and the 'chemo air pharmacology program'.

Impairment

After close-up observation of the effects of Macondo smoke and toxins downwind in Arkansas, with flash floods caused by the soot seeding clouds over the Caddo Valley and an HAAD aerial, display

greater than any seen by myself and colleagues.

We concluded that the dangerous levels of UV were close to dropping people and more birds in their flight.

The total accumulated BP effects on the ozone had sicken the total life-support system. The low and high level massive spraying was an attempt to fill the hole in the sky burned by BP.

This reality and the effects on me and dogs was when a heavy straw fell.

I had enough and with pounding head I drove south and soon the headache lessoned as I got away from the Little Rock toxic cloud.

Leaving one mind numbing sky mess I Returned to New Orleans before November 26, 2010. The final day for emergency payouts from Feinberg.

The day the lawyers began to stall me. I was in the news, hell, Michael Moore even had me on his website front page for months.

I had not been able to think out what had happened to me and many others completely.

One symptom of Btex or toluene and benzene poisoning is impairment of mental functions: A common complaint in St Bernard Parish where 90% of the toxins passed through in the wind and in the tidal waters. Here was where 20% of people tested, as report on New Orleans Channel 8, showed high levels of toxins in blood and tissues traceable to the spill and its aftermath.

After some luck and a lot of study of the available information, and over 100 hair sample from St Bernard fishermen and families and other 'Downwinders' and a few upwind in the keys, I discovered what was making me and so many others sick and tired along the coast and many places around the world.

After seeing several of the hair sample lab results, well know Dallas surgeon and environmental medicine specialist, Dr. William Rea said, with great surprise, "where did the uranium come from".

Many were near lethally contaminated with this radioactive material. This is a very important question that we will attempt to answer.

In my information search I discovered that there is a mathematical and chemical process called inverse

function or inverse proportion.

This process was put in motion in the body chemistry of all mammals and humans, including me.

For many months, overwhelming toxins were in the air, and measured by the EPA, creating a negative environmental condition. It is affecting the health of all residences and mammals around the Gulf Coast and many miles downwind of the SPILL event today July 2019.

The Process of Disease Invasion of Mammals.

This process involves a mathematical relationship in which an increase in one variable causes a decrease in another variable or biochemical factor.

In my case and that of all others living and working along the affected coast, and upwind hundreds of

miles, this variable was toxic metals and solvents in the life support system.

The given factor was the SPILL event. This process brings about a decrease by the same factor in another entity.

There is a clear cause-and-effect relationship in the illnesses suffered by humans in the communities along the Gulf Coast in the aftermath of the SPILL and clean-up activities that followed.

In other words: This mathematical operation or function of specific light and heavy molecular toxins in the air, coming from the BP Macondo well and other closer sources, and water exactly reverses or displaces another operation or function.

The release of dispersants and other chemicals upwind of populated areas led to the displacement of atomic-weight molecules of essential elements from the bodies of mammals and humans that were exposed to molecularly-heavier heavy-metal toxins.

This is a primary cause of the declining health of the people and other mammals living in coastal communities and in the waters, along

the shore who were affected by the spill, including myself.

After much consideration and consultation with prominent internal-medicine physicians and with alternative-care doctors, nutritionists, and health-care providers, in environmental and endocrinal medicine and physicists, my and their conclusions are simple: In the long term, those exposed and still living along the coast will not regain stable health without proper chemical detoxification.

This is needed to remove these toxic metals in everyone's body that resulted from the Spill and "normal" poisoning by the system.

What is considered normal in today's world will assure disease, and without proper replacement of essential trace metals and beneficial minerals through replacement therapy all are being effected Daily negatively.

This displacement of essential minerals and nutrients is the cause in part of many medical complaints and re-mineralization of each affected

individual is one necessary step to improve community health. These high levels of toxins were present in the environment, in which I and so many innocent, and a few guilty, hard-working people lived and worked, as a direct result of the Spill and an ultra-toxic to the wind chemical and refinery industry.

Louisiana oil and industry is totally controlled now by multinational companies with roots in the Germanic 1000 business plan.

No one in New Orleans, or and especially, St. Bernard Parish could feel good with the levels of multiple toxic chemicals in their bodies as revealed in recent health studies.

My one of a kind, open data base hair sampling exposes exposed a grim reality for all of us. Look at Public Eyes Media.org, our site. There you can order hair sample test and share your results in an open data base hair and read an Alzheimer's and cancer recovery testimony from a prominent phycologist.

Arkansas got the downwind hit as shown in my film Vitamin D Deception seen in many internet places and on Free Speech TV. See in on Geo-Engineering Watch/vitamin D deception.

Rigging the Game

One of the hurdles set up by the BP fund lawyers was the requirement that any attending physician had to make the determination that the spill was the cause of health issues.

My New Orleans doctor, Dr. Ott, is a Primary Care Physician and thus he is not qualified to state, as a toxicologist would, that the spray was the cause of my illness but my proximity to the toxins in the smoke and from the spraying at close range, upwind of our boats and homes, gave a good idea of the cause and effect relationship.

He did state the information that I was at the scene and nearer the well then most and I came to him for treatment for what I knew was caused by my extreme exposures.

Everyone around me including all crewmembers on our two boats. All, but one, had to seek medical help and two ended up having their hearts unplugged from chemically released fats that were dissolved by the oil dissolving dispersants clouds.

For me, and the entire coastal community, toxic poisoning and medical and physical injury reality there is no lack of documentation: There is more than enough data establishing a cause-and-effect relationship with the Spill, the dispersants, and coast-wide and downwind health issues beginning in May 2010.

But there is no wind in the now BP smoke and chemical saturated stratosphere where damage will continue over the region until it is no longer even remotely safe to go outside during the day and maybe in the night.

Per EPA data collected during the SPILL and its aftermath, everyone should have left the New Orleans area and abandon their work for months if not years.

Everyone should have been evacuated from many coastal communities if not for the agenda and the formula for not helping people leave the collateral damage zone.

This was in effect and many were assumed to die as collateral damage as a cost of eco-terrorism, in the aftermath.

Death, and whole family illnesses, is rapid on the bayou since Katrina and now this sucker punch is designed to finish the job of depopulation and resource control grab. Just like in war but the silent war of the NWO cowards and the bankers.

After two years of gathering data and following the reality of life in the bayou, as one attorney in NO commented, "Something is really different with people; I see it more and more clearly each time I go down the road."
A Cajun term for going south toward the sea.

I have a research paper trail, included in this book that should forestall any further need for more research to prove the factual medical conclusions that validate the Macondo health aftermath situation.

I, like so many others who worked and lived close to or downwind of the Spill, are now chronically affected and most are chronically fatigued with toxin-induced adrenal fatigue.

Minerals and vitamin d plus EDTA oral chelation has worked for me.

Many are deteriorating rapidly and want medical help to get their health back. Most people that I know in St. Bernard Parish believe that medical health is the most important factor in their difficult lives.

I have spent a good part of the last nine years trying to find out clearly what was wrong with me by gathering official documentation of the spill and obtaining medical opinions, diagnosing and researching treatments that safely remove toxins from our bodies.

This included the bodies of those young men and women who worked in the clean-up process.

They were the only ones protecting the marshes and waterways and the seafood those habitats provided.

These people are your gods for they give your life substance. These fishermen, crabbers, shrimpers, oysterman are essential to seafood consuming citizens.

Sickened fishermen, farmers committing suicide at astounding rates, and they are the ones who feed you and give your life fuel to live. Period

All along the coast, and far inland, man and beast

experienced, involuntary exposures to extreme levels of toxic metals, volatile organic compounds, radioactive oil smoke, and toxic oil smoke.

BP and our collusive government released this into the coastal environment and to the winds, without proper warning of the magnitude of the event and reasonable evacuation precautions or proper safety equipment.

My physical injury from toxic poisoning caused by the spill was sustained while I was performing activities authorized by Parish government officials, gathering oil spill information and resource damage assessment.

I had been living in St. Bernard Parish and working where levels of Spill-related toxins above legal limits were measurable daily for the many months of daily and hourly air tests.

The toxins were still present in measurable levels when the testing was discontinued and I was advised by my doctors (3) to stay away from my work area and to stay inside when home due to the toxics effects on my body and mind.

High levels of SPILL toxins were continually measured and detected in uptown New Orleans all throughout 2010 and well into 2011 and still no health warning.

The wind and rain streams of measurable high levels of toxins from the Spill were always there and the only thing that changed was the wind direction.

The danger of exposure in St Bernard Parish and in a lesser degree New Orleans will go on for many years. The oil is still bubbling up and more sea life is showing ill health in the ecosystem at extinction event degree.

At no time were any of the boat crews or myself offered respirators or given warning that toxic levels were well above levels that should have required evacuation.

My illness/poisoning were among the first reported by local and worldwide media.

At the time of the exposures, I was staying at a house owned by George Barisch, President of United Commercial Fishermen Association.

At the insistence of Mr. Barisch, the news media were called in May 2010 because of the severity of my sickness.

He felt the need to warn those working near the toxins to protect themselves and to inform the response organizations to better equip their workers.

My exposure and first serious illness took place in early May 2010 before formation of the GCCF Fund.

My lungs have never completely cleared for months and any trip that I make to the bayou or to other areas near my bayou home brings on lung congestion and coughing spasms.

I have become so chemically sensitized that even a drive by a smoking refinery affects my health.

High Levels of Toxins in Everyone and their dogs

The presence of many toxic materials, above any background or normal pollution levels, has been

verified by several governmental and private labs by testing human tissue samples, air, water, and rainwater of the area.

In my tissue sample, 17 poisons were identified in high levels. This was the average number of toxins in all hair samples we took and had analyzed.

The EPA recently released the list of 150 ingredients in Nalco's 'Corexist' dispersants. Given the presence of numerous toxins in my body from the Spill, it is accurate and confirmed that they are present in my body at toxic levels.

Most toxins revealed in the EPA Air Sampling Reports and Volatile Organic Compounds Reports had a direct chemical connection to the spill, the dispersants, and the aftermath of the burning and sinking of the Spill.

The poisons entered everyone following the Spill via the air we breathe, the water we drink, and our food, and clearly by just living near the coast and on the water.
Many spill-related toxins are still in the air along the coast and over Arkansas. This is due to sunken

oil continuing to leach up from the bottom with the rise in water temperature and storm-related wind and wave action.

July 2019 Tropical Storm Mary will un-lease a toxic windstorm and rain of sunken BP oil toxics that will hit Arkansas and other states, with near lethal contamination of rivers, streams and ground water. Drink a little BP tomorrow.

Unknown to me for weeks was the existence of an EPA VOLATILE ORGANIC COMPOUNDS AND EPA AIR SAMPLING monitoring device that was located just blocks from my residence in Arabi, LA.

After research into the EPA gas monitoring, we discovered that the Arabi, LA. site monitored gases passing through St. Bernard Parish blowing from the south or from the BP Macondo well, on the way to downtown New Orleans and north into Arkansas and other states.

The EPA reported and the press passed on that "toxic levels of Benzene, Ethyl benzene, Toluene, and Xylene were in the air. These blood thinning solvents were in the air we had to breath for months.

The local press reported often, during the summer and fall of 2010, that on some days the levels rose mysteriously hundreds of times higher than any known recorded background.

Way above the Spill 'norm' for crude oil composition established by these daily air and water tests.

Toxic aluminum and nickel were reported in rain and in human tissue samples. All these profound warnings should logically have required an evacuation and warning of questionable water safety in Arkansas.

A secret to state officials testing of material in the possession of the state labs, more than ten years ago, of tissue of fish from every major stream in Arkansas. All showed Aluminum, at refined nano particle size, not naturally occurring aluminum.

This material is common debris from all space craft using the USA formula for rocket fuel.

This "natural" occurring aluminum is used every day as the cover story to cover the aluminum contamination caused by NASA/NAZA

Several independent health studies were conducted during the last few months that analyzed human blood, urine, and hair samples.

I participated along with Vessel of Opportunity crews, and their families. Toxic chemicals, traceable to the Spill in measurable amounts, were found in samples from everyone tested.

Some of the toxins were at extraordinarily elevated levels. They were far above any lifetime body burden resulting from normal pollution.

Early in the Spill the EPA released data that Benzene levels in New Orleans had rocketed to as high as 3,000 parts per billion.

Benzene is extremely toxic, even short term exposure at low levels can cause agonizing illness and slow death from cancerous lesions and leukemia years later. But 3,000 ppb is far from a low reading.

Hydrogen Sulfide was also detected by the EPA monitoring stations around the New Orleans area. The EPA reported Hydrogen Sulfide levels as high as a 1200ppb periodically for weeks. A normal safe level falls between 5 to 1 ppb.

Chronic Fatigue

To help validate the health effects of the Spill on my health and of many others exposed, I enlisted the advice of several specialists in environmental medicine and toxic-exposure symptoms.

After viewing the test results and hearing the complaints and reading the health studies, their conclusion was that the presence of 13 to 15 neurotoxins and solvent compounds most likely could be the cause of various medical issues and could explain the chronic fatigue plaguing St. Bernard Parish residents both young and old.

One study found 20% of coastal residents with residues of dispersants and crude oil related toxins in their blood.

My residence during the Spill was in Arabi, LA. until I was forced to leave the contamination area on my doctors' advice.

EPA data showed there were days beginning in early May 2010 until March 2011, when the already high levels of many toxins, particulates, and VOCs jumped to the previous extremely elevated levels.

Residents depended on governmental agencies for good judgment regarding toxic chemical levels in St. Bernard Parish, New Orleans, and areas to the North. We had no way to judge the danger we were in, far from the Spill, since nothing of the magnitude of this event had occurred before.

It seemed that no one understood the carrying capacity of the winds to hold aloft and move toxic pollutants inland. Nothing of this magnitude had ever openly happened, in our knowledge, except for the Ixtoc-1 oil spill in the Bay of Campeche in Mexico in 1979 and Mt. Pinatubo volcanic

We had no frame of reference for millions of gallons of burning oil and chemicals and the effects of hundreds of billions of cubic feet of natural gas and BETX compounds (Benzene, Ethyl benzene, Toluene, and Xylene), and many more VOCs. (total 57 Volatile Organic Compounds present) in our air supply nor their effects on humans.

Those hydrocarbon compounds were toxic to humans and other mammals including Bottlenose Dolphins and Sea Turtles that could not escape from the spill zone. Many species of sea life appeared disoriented and dazed during and

following the Spill. Brown Pelicans became comatose to the point of non-resistance. They just laid still along the shoreline or in shallow, oil-filled waters.

During the event, there was no place inside or out where humans could get toxic-free air to breathe.

A science and toxicology report on dispersant fumes stated one of the first effects of toluene and other BTEX compounds would be to quickly render olfactory nerves useless.

Because of this nose- numbing effect, the danger of VOCs in the air was not evident to humans after the Spill and continued for hours, and days, and weeks, and months. Our fight or flight response was rendered inoperative due to toxic effect.

Humans and animals alike were poisoned and debilitated. Many St. Bernard Parish residents, and many of my friends developed severe headaches, especially those living near the oil-impact zone that received VOCs from the continued volatilization from floating and submerged deposits of oil.

Young men whom I have known for a few years are getting old and sick right before my eyes.

My most disturbing symptom, now common among coastal residents, was nosebleeds. For a ten-day period in March, 2011 I experienced long and tiring nosebleeds every day.

One bleeding incident happened at a registered nurse's residence who immediately checked my blood pressure and found it was normal.

I was finally advised to leave the heavily polluted coast for a while.

I called Tokyo Broadcast, whom I worked with as a field producer/stringer and a crew came the next day.

We broke the story of the wall of booms between Mississippi and Louisiana that broke the oil into a thin layer of oil. Less of a fine or no fine. This story was blocked in America.

BP with our government in tow got away with environmental disaster for not one cent after tax breaks in the dark.

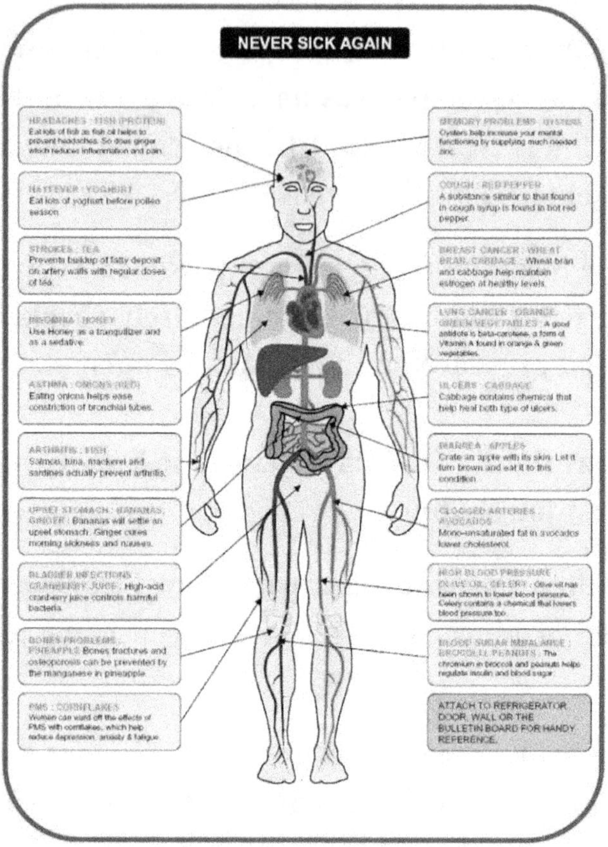
NEVER SICK AGAIN
HEADACHES : FISH (PROTEIN)
Eat lots of fish as fish oil helps to prevent headaches. So does ginger which reduces inflammation and pain.
Eat lots of yoghurt before pollen season.
STROKES : TEA
Prevents buildup of fatty deposit on artery walls with regular doses of tea.
Use Honey as a tranquilizer and as a sedative.
Eating onions helps ease constriction of bronchial tubes.
ARTHRITIS : FISH
Salmon, tuna, mackerel and sardines actually prevent arthritis.
UPSET STOMACH : BANANAS, GINGER : Bananas will settle an upset stomach. Ginger cures morning sickness and nausea.
BLADDER INFECTIONS : CRANBERRY JUICE : High-acid cranberry juice controls harmful bacteria.
Bones fractures and osteoporosis can be prevented by the manganese in pineapple.
PMS : CORNFLAKES
Women can ward off the effects of PMS with cornflakes, which help reduce depression, anxiety & fatigue.
MEMORY PROBLEMS : OYSTERS
Oysters help increase your mental functioning by supplying much needed zinc.
COUGH : RED PEPPER
A substance similar to that found in cough syrup is found in hot red pepper.
BREAST CANCER : WHEAT BRAN, CABBAGE : Wheat bran and cabbage help maintain estrogen at healthy levels.
LUNG CANCER : ORANGE, GREEN VEGETABLES : A good antidote is beta-carotene, a form of Vitamin A found in orange & green vegetables.
ULCERS : CABBAGE
Cabbage contains chemical that help heal both type of ulcers.
Grate an apple with its skin. Let it turn brown and eat it to this condition.
CLOGGED ARTERIES :
Mono-unsaturated fat in avocados lower cholesterol.
HIGH BLOOD PRESSURE : OLIVE OIL, CELERY : Olive oil has been shown to lower blood pressure. Celery contains a chemical that lowers blood pressure too.
BLOOD SUGAR IMBALANCE : BROCCOLI, PEANUTS : The chromium in broccoli and peanuts helps regulate insulin and blood sugar.
ATTACH TO REFRIGERATOR DOOR, WALL OR THE BULLETIN BOARD FOR HANDY REFERENCE.

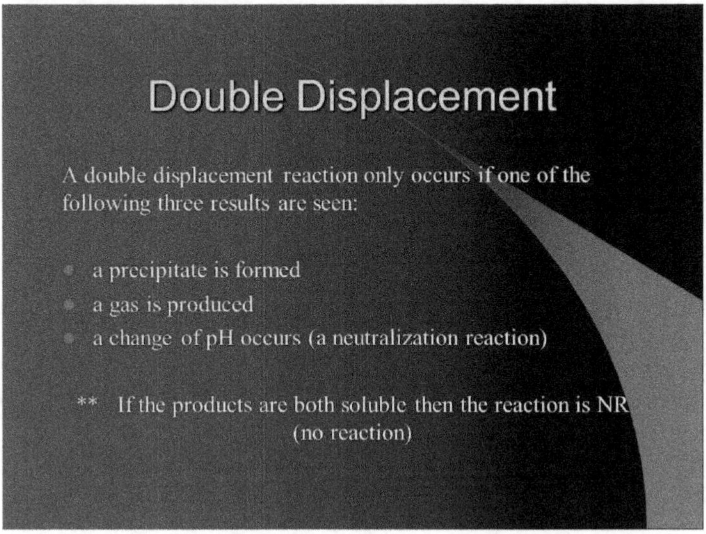
Double Displacement
A double displacement reaction only occurs if one of the following three results are seen:
a precipitate is formed
a gas is produced
a change of pH occurs (a neutralization reaction)
** If the products are both soluble then the reaction is NR (no reaction)

THAT CRACKING WHEN THE CHIROPRACOR ADJUST YOU OR YOU CRACK YOUR KNUKLES. THIS IS CO2 GAS THAT HAS FORMED IN THE BODY IN EXCESS DUE TO THE 31 PERCENT IN CO2 BEING BREITHED IN. THIS IS THE AMOUNT OF ATMOSPERIC INCRESE SINCE 1950. THIS IS ANOTHER POINT IN MY THEORY OF A GLOBAL CARBON DISEASE.

***

Neutralization Reactions

Is a special type of double displacement reaction where an acid reacts with a base to produce an ionic salt and water

Acid + Base → Ionic Salt + Water

- Acid: $H^+_{(aq)}$ and $X^-_{(aq)}$. $HNO_{3\,(aq)} \rightarrow H^+_{(aq)} + NO_{3\,(aq)}^-$
- Base: $M^+_{(aq)}$ and $OH^-_{(aq)}$. $NaOH_{(aq)} \rightarrow Na^+_{(aq)} + OH^-_{(aq)}$

$$\mathbf{HX_{(aq)} + MOH_{(aq)} \rightarrow MX_{(aq)} + H_2O_{(l)}}$$

$$HNO_{3\,(aq)} + NaOH_{(aq)} \rightarrow NaNO_{3\,(aq)} + H_2O_{(l)}$$

If I had been a chemist I would have known a long time ago what they were doing to us the HUMAN cattle on their planet.

Below is part of the answer of intentional toxic contamination to fill hospitals

Displacement reaction

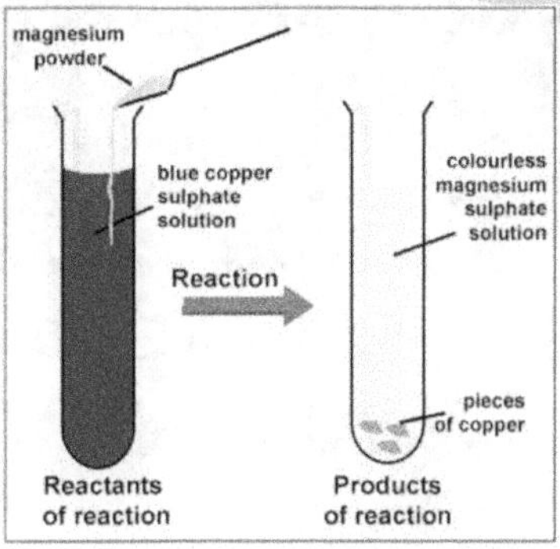

Displacement reactions occur when a metal and salt (metal + non-metal ionic compound) solution are mixed and the more reactive metal replaces the metal in the salt.

An example would be reacting magnesium metal and copper sulfate to produce magnesium sulfate plus copper metal.

These are limited to the displacement of metal ions in solution by other metals.

3. Single Displacement (cheater)

One element knocks out ("displaces") another element in a compound

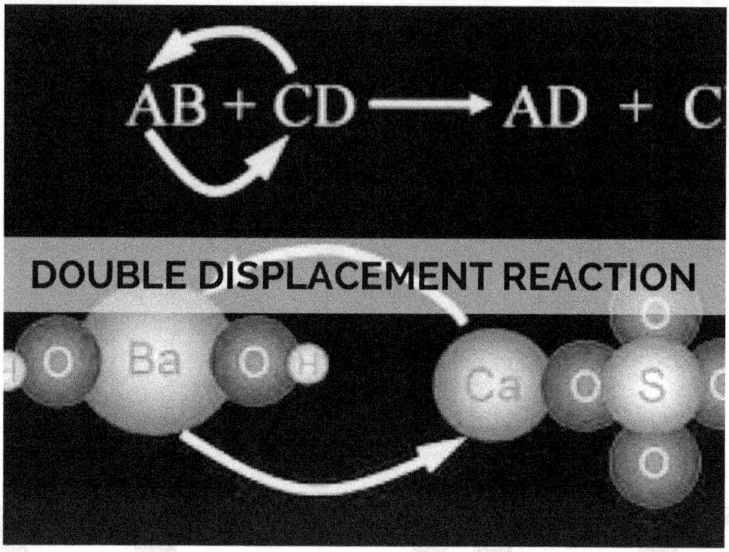

AB + CD ⟶ AD + C
DOUBLE DISPLACEMENT REACTION
O
Ba
O
H
Ca
O
S
O
O

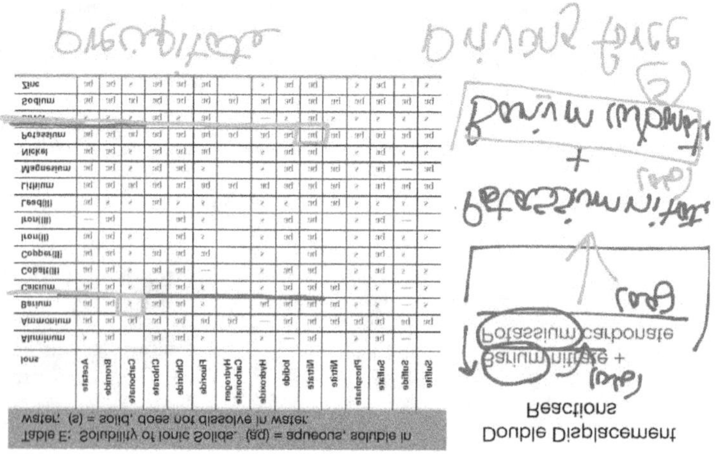

Double Displacement Reactions
Barium nitrate +
Potassium carbonate
Table E: Solubility of Ionic Solids. (aq) = aqueous, soluble in water. (s) = solid, does not dissolve in water.
Ions
Aluminium
Ammonium
Barium
Calcium
Cobalt(II)
Copper(II)
Iron(II)
Iron(III)
Lead(II)
Lithium
Magnesium
Nickel
Potassium
Silver
Sodium
Zinc
Acetate
Bromide
Carbonate
Chlorate
Chloride
Fluoride
Hydrogen Carbonate
Hydroxide
Iodide
Nitrate
Nitrite
Phosphate
Sulfate
Sulfide
Sulfite

WE MIGHT HAVE ENOUGH TIME.

BUT WE MUST STOP THIS CORPORATISM

DO NOT WORK FOR THEM

AMERICAN BUSINESS NAZI's WITH TOTAL BUSINESS CONTROL OVER OUR LIFE SUPPORT SYSTEM, INFILTRATED WITH HELP FROM WALL STREET MUST GO.

BANKERS OF GERMAN DESENT ARE STILL HOLDING

AMERICA AS AN ECONOMIC HOSTAGE. THIS MUST STOP.

THESE BANKERS CAPTURED THE ECONOMY AND SOUL OF OUR COUNTRY. AGAIN, THEY WERE BROUGHT TO AMERICA AT THE END OF WW2.

GERMAN/AMERICAN BANKERS

AND PLANNERS ARE THE ONES DRIVING THE CONTINUED INFILTRATION, OVERTAKING, AND UNDERMINING AND

POISIONING OF THE AMERICAN IDEA OF PERSONAL FREEDOMS.

DEMOCRACY IS AN ILLUSION BUT A GOOD ONE TO TRY TO MATERIALIZE

THEY ARE IN POWER TODAY. THE BANKERS FUND AND RECEIVE PROFITS FROM ALL SIDES OF ALL WAR BY FUNDING BOTH SIDES. WALL STREET IS A RIGGED GAME THAT FUNDS THIS GAME.

STOP BUILDING BOMBS FOR

THE STUPID GENERALS

FAIRISM IS THE NEXT FORM OF GOVERNMENT OR ELSE

STILL HOT ATOM BOMB TEST SITE IN AMERICA

AND A DUMB TOURIST ATTRACTION

Senator Johnson supported Von Braun, the man behind killing Kennedy, to continue Vietnam and the infiltration. He had embraced the madness.

"We shall propose further cooperative efforts between all nations in weather prediction and eventually in weather control."
- President John F. Kennedy Sept. 25, 1961
WEATHER WARFARE

KENNEDY WAS AN ELITE, HE SPENT A TIME IN GERMANY IN LATE 1930s GOING THROUGH INDOCTRINATION. HE WAS ONE OF THEM. THEY PROMISED HIM THE PRESIDENCY UNTIL HE REALIZED THE REAL GOAL. THEY HAD TO KILL HIM.

Why?

They possess a secret common goal

of world dominion and central control

of the manufactured system and control of minds as slavery

Not a bad idea in a world of confusing ideologies

SO WHY KEEP IT SECRET FROM THE PEOPLE IN GENERAL. WELL, IT IS EASIER TO BE SILENT THAN TELL THE TRUTH WHEN YOUR JOB DEPENDS ON NO ONE KNOWING MUCH.

Damage is done so what can one do now to buy time and keep quality of life longer

GOLD HEALS

The story of gold is a central thought needed to understand why gold is greed evoking and why gold is something to be coveted and kept hidden away and something to kill for.

Why, from land dissipated by thousands of years of farming, and over farming, come the ones with gold fever.

These killers roamed the world looking for gold, an essential nutrient, without understanding the central thought that gold is meant to flow through the planet body over great time and through the living bodies as a molecular regulator.

When the gold in the body is also decimated and dissipated by lack of uptake the gold fever ensues and consumes weaker minds

The paradox of those that possess gold and do not consume it and those who take it and spend but never put it back into the soil from which we grow food.

We all depend on this mineral and nutrient substance for mental health. A healthy golden soil

like those of the Aztecs and Mayan gave freely through the food chain is the key to enlightenments.

When their soil was depleted they declined.

There is a golden peaceful feeling that consuming gold gives one.

Gold is also the trump card in the immune system , without it lower base metals and thoughts rule.

The modern replacement for gold is fluoride and chlorine atoms. These brain toxins, with electrical energy like microwaves, allow the control of moods and the lives of those effect by mood swings.

We are affected everyday by swinging moods caused in part by the lack of gold in our life chain of atoms.

The ancient Ayurveda text states: "Gold is soothing, pure, nutritive, curer of poison, phthisis, insanity and other diseases.

"Gold increases vitality, fortune, beauty, intelligence and memory. It destroys all sorts of diseases, pacifies the evil influences exerted on human beings by ghosts, is an aphrodisiac and gives rise to happiness and nutrition. It cures disease, prevents senility,

removes loss of memory and consciousness. It also removes thinness, develops the mind, and increases semen."

Gold reverses the effects of fluoride poisoning in those who drink fluoride water.

Vaidya Bhagwan Dash writes in his contemporary work on Ayurveda: "The bhasma of gold is sweet in taste and vipaka (taste that emerges after digestion and during metabolism). It is aphrodisiac, cardiac tonic, promoter of eyesight as well as intellect and rejuvenating. It counteracts the toxic poisons. It promotes the complexion of the skin." Under the heading Indications he writes: "It promotes longevity, maintains youth and memory. It cures serious types of fever, particularly chronic fever, nervous disorders, heart disease, tuberculosis, afflictions of voice, schizophrenia, epilepsy, hysteria, bronchitis, asthma, chronic diarrhea, serious types of anemia and cancer."

For those who engage in meditation practices, it is a tool to reach expanded states of consciousness and sublime states of mind easily and effortlessly.

In Alchemy, we see gold as the representative of the sun, which rules the fourth seal or chakra of

the human energy body and rules the heart and the arteries, as well as the skeletal structure of the human body. Based upon this knowledge, it is not surprising that we can observe the following effects of gold: On a physical level, monatomic gold increases stamina. Monatomic gold increases and balances the production of your own hormones, thus being useful for rejuvenation. Monatomic gold strengthens the heart, and increases the production of red blood cells. Monatomic gold is an aphrodisiac for both genders and increases the production of semen in males.

Earth ruled by defective DNA Dominators. They are kinky and defective but in control of the money

The lower DNA politicians, Military generals, admirals and those in authority are gold deficient and are on guard against potential peace breaking out.

These military minded are attracted to glitter like lower level primates or like a woman wanting a shiny wedding ring.

The cause of the underlying desire to capture and trap another DNA group into servitude is the nesting response of females of all mammals.

Finding a home to lay eggs or produce babies is a primary driver of conquest. My theory is simple: through breeding programs, or arranged wedding, selective traits are brought into dominance. Depending on the area of the world one was

spawned and born certain metals and nutrients- metals caused hormonal changes in mammals. Mineral depravation was mind altering and maddening. Men became over extended into the realm of taking and killing to provide nesting materials and food to continue the DNA dominance invasion. Communism and other totalitarian government existed to force communal sharing and

"Farism".

Problem was all leaders became overseers and they lost to the mineral madness and desired dictatorial powers. Never fails. And Trump is obvious example.

The earliest example came from the Kazarian, a future Germanic, culture of the mongrel's. They are the predominate aggressive DNA infesting Earth. The basis cause of lack of MORAL evolution.

My impression after seeing generals and dictators, like Amin, wearing a breast full of metals IS these are very low DNA strain people. More shiny metals

more likely to mass murder and get higher pay for killing.

Those brilliant scientists like the fifth columnist Einstein were math functional and moral deficient and ego driven to show their stuff and gamble on our future.

They were lower DNA and the making of mass murder and planet killing bombs was ok if you spelled their name correct and praise them for brilliance and madness.

This is the reason Earth is dying. Atomic bombs in the air, under and over the sea and over cities and sell us that this is progress.

All people in high positions seem DNA defective. The British crown is so inbred that they recently brought a strong blackish strain of new DNA to the crown to head the DNA disaster they're part of.

Queen Victoria was the grandmother of all European monarchies during the 19th century. Her defective DNA brought hemophilia into the inbred bloodlines across the world. This DNA defective woman changed the planet in a natural way.

An answer could be found in logic; in reason; in a moral standard balanced just like the energy releases from our sun that become mammals; become all things.

So, all things equal and fair I have proposed for a decade now the idea of fairism. Fairism views all resources on the planet as equally shared and provide equal to all creatures. So, dividing the wealth, including oxygen, as equal property and subject to the use permission of the end users that includes all mammals.

Example of a right mind thinking: Would you consent to giving a million life times of air to a single jet liner flying with three passengers—this happened to me—that reduces, the potential of the planet surviving until evolution is complete and the planet dies naturally.

I am certain that, given the validly of my story of life on Earth in big trouble, I was given the insight of telling a little at a time and offend few of the guilty—but times of universal deceit are over the evidence is immediate and if our time is almost up then "all is in the all".

All Truth Passes Through Three Stages.

First, it is ridiculed, second it is violently opposed, and third, it is accepted as self-evident. – Arthur Schopenhauer Philosopher, 1788-1860

The terminal problem on earth now is simple: A much lower strain of DNA has occupied Earth since Adam and Eve time. Since this earth life incarnations beginning. The evidence is simple and profound.

The lower DNA invaders are programmed to kill and to develop and encourage the use of potentially planet killing weapons that in time will kill the host organism.

These low lifers are like cancer, when attacked, goes into hiding and returns as secondary cancer more than 90% of the time.

A simple requirement by oncologists for all cancer patients is to chelate and detox from the chemo after effects-if you survive. Immediate and constantly re-materialization with essential minerals and micro minerals.

A prime essential nutrient on this planet is gold

We are here for a reason this book

is one of mine

Scientists or the ones who are close to the reality I have brought forth will be relieved that I did it. Took a while. If I was a chemist I would have figured this out about a minute into basic chemistry. All through school I had a job so learning my work schedules was my priority during junior and high school.

The point I make is: Science knows the reality in the number. Most never know enough to put a face on a fact and disease caused by their learned facts.

Some scientists have discovered that living in an acid environment and having an acid input, like excess CO_2 we breath every second promotes

"DEMINERALIZATION" leading to osteoporosis and heart disease. This disease is spreading like the flow of excess co2. Which had risen 5 ppm in the months of May and June 2019.

The tipping point has tipped.

An acid world atmosphere state like we are in now effects the thyroid and this affects all organs and enzymes will not bind properly.

Action time was back when the first guy in England, over 120 years ago, told us this was happening between the lines of co2 warning.

CO2 is effecting our chemical equilibrium. Not good like having bad gas and your car is running rough.

Kidney stones of aluminum and carbonic acid are now being removed from older patients who live inside with higher levels of co2 in the air inside. (see public eyes media.org. (Please read the letter from Doctor Carroll)

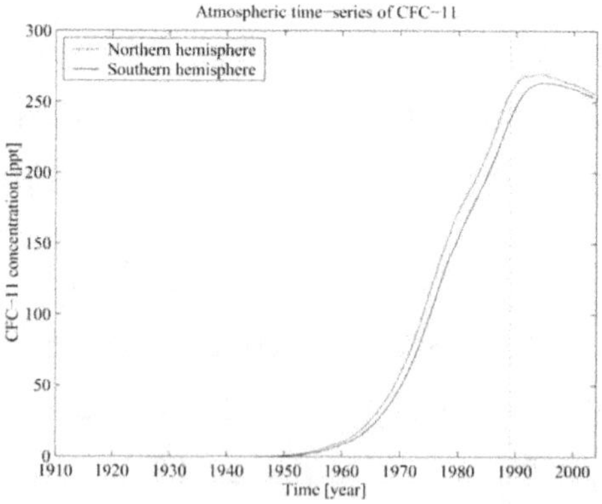

1950 AGAIN

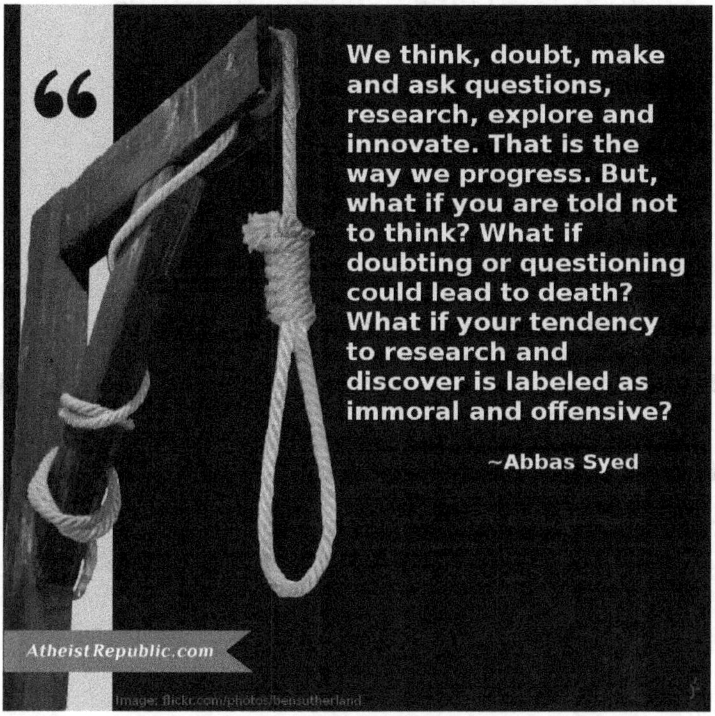

66
We think, doubt, make and ask questions, research, explore and innovate. That is the way we progress. But, what if you are told not to think? What if doubting or questioning could lead to death? What if your tendency to research and discover is labeled as immoral and offensive?
~Abbas Syed
AtheistRepublic.com
Image: flickr.com/photos/bensutherland

I repeat

Aluminum, Copper, Mercury, Lead, Cadmium, Arsenic Toxicity Test

An average person that eats healthy and lives a healthy lifestyle can still be suffering from toxic metals burdening their body. When metals enter the body, they get stored in the soft and hard tissues. Since your hair is the 2nd largest tissue in the body it is an effective way to test the person's overall metal toxicity.

When a person has an overload of metals their body has a harder time healing and performing optimally.

Aluminum Toxicity: Aluminum is a nonessential element that can be toxic if excessively assimilated into cells. Aluminum can lead to symptoms of brain fog, dementia, poor behavior, learning disorders like ADHD, autism, renal problems, headaches, pain syndromes and more.

Cadmium Toxicity: Cadmium has no metabolic function in our bodies. High levels can contribute to symptoms of hypertension, hypotension, fatigue, weight loss, osteomalacia and lumbar pain. Cadmium can also cause problems with the kidneys, lungs, arterial walls, bones.

Lead Toxicity: Lead is a neurotoxin which inhibits the body's ability to utilize the essential elements calcium, magnesium, and zinc. At high levels Lead can affect the memory, cognitive abilities, nerve conduction, learning in children and more. Lead toxicity can contribute to loss of appetite, weight loss, poor memory, fatigue, constipation, headaches, brain fog, decreased coordination, and more.

Copper Toxicity: High levels of Copper can contribute to liver disease, and renal dysfunction. Some of the symptoms of high Copper levels are muscle and joint pain, depression, irritability, tremor, hemolytic anemia, learning disabilities and erratic or moody behavior.

Mercury Toxicity: Numbness, weakness of the legs, spastic paralysis and impaired vision are just a few symptoms of toxic mercury. Mercury exposure can come from the environment, dental amalgams and seafood. Some symptoms of mercury poisoning are brain fog, Alzheimer's symptoms, kidney problems, infertility, PCOS, pain syndromes, insomnia, headaches, tremors, lack of coordination, blurred vision, neurotransmitter imbalances, MS (Multiple sclerosis) and immune challenges.

Once Mercury gets into the central nervous system it can contribute to emotional, psychological and neurological problems.

Every time we chew, brush our teeth, or drink hot liquids, mercury vapors are released, can cross into the blood brain barrier and start causing problems.

Neutralization Reactions

Is a special type of double displacement reaction where an acid reacts with a base to produce an ionic salt and water

Acid + Base → Ionic Salt + Water

- Acid: $H^+_{(aq)}$ and $X^-_{(aq)}$. $HNO_{3\,(aq)} \rightarrow H^+_{(aq)} + NO_{3\,(aq)}^-$
- Base: $M^+_{(aq)}$ and $OH^-_{(aq)}$. $NaOH_{(aq)} \rightarrow Na^+_{(aq)} + OH^-_{(aq)}$

$$HX_{(aq)} + MOH_{(aq)} \rightarrow MX_{(aq)} + H_2O_{(l)}$$

$$HNO_{3\,(aq)} + NaOH_{(aq)} \rightarrow NaNO_{3\,(aq)} + H_2O_{(l)}$$

DRUGS AS A TOOL OF INVASION

Tonight, on PBS May 18, 2019 the story of the mad British drug dealers and controllers addicting millions as an invasion tool.

Almost all mass-produced drugs, like meth, have some connection to the not so royal crowns by their ownership of the companies providing the base chemicals to make the invasion tool.

The crown power also provided immunity from prosecution and exposure. Control the information and win hearts and minds was the plan.

Drugs have been used to control or numb minds and bodies since our beginning.

It seems man has a flaw: he consumes toxic plants like tobacco and turns food plants into alcohol and mass misery through mental illness caused by the toxic chemical alcohol.

The warped world mind, in control takes beneficial plants like, opium and misuses it for profit and pleasure not relief from pain.

Advantage over people has been key to the organized misuse and mind control antics of control freaks across the planet for a long time.

Old English witches brew contained numerous psychoactive compound. These witch's brews could give the witch or power manipulator the power of suggestion over the minds of the masses. Mass media uses the powers of suggestion everyday with drug ads

The planets' most prolific conquistadors of the past, present and future, the British and the Dutch Crowns invaded China with opium and two wars were fought between 1839 to 1860 to run the invaders from the land.

The objective: drug/chemical slavery of China and in time the planet. Today meth, coke, heroin and most large scale drug slavery is still rooted in these inbreed monarchies desire to control everyone.

Thank god for hemophilia which killed off most of the inbreed crowns of Europe.

Russia was taken down in part by hemophilia in the Romanov royal family.

Out of these English and Germanic monarchy and other ego driven society controllers came the World Wars and a New World Order plan. The use of mind and body control drugs like saltpeter were commonly used by military and prison officials and in the 1930s' Hitler ordered drugs be invented for a super solder drug. Methadone was the big one and it is used to enslave the very weakest in American Society today. Meth was standard medicine before any battle.

Out of 1930s' Germany and on order from Hitler new drugs were invented for a war purpose and later continued drug invasion of easily weaken countries like America.

The weaken state of America comes directly from the use of mind numbing drugs like fluoride and metals like aluminum, barium, cadmium. These metals all have drug like effects on the mind. The further development of invasion drugs like fentanyl continues and kills many of the weakest in society. Kind of feels right at times. Culling the weak minded maybe a good thing, maybe not. Most creative writers used alcohol and drugs to create so the dilemma is clear to me.

An example of a pain drug designed to control the minds of those who over-consume is OxyContin.

This comes from genetically modifying poppy for high levels of thebane.

This genetic modification invades minds and in many case, create damage from which recovery is less likely.

This drug invasion evidence of intentionally addicting is now in the news with the Sandler family, finally a drug invasion society weakening family may be going to prison.

I must wonder are "they" part of the "conspiracy" that is drugging white folk in countries with fluoride as well. Yes, blacks are not the only ones going down. Lot of white folks dying of oxycodone, blacks die of other drugs more often.

The question of intentional addicting is answered simply. Yes. This is the character of dominion planners. These drugs displace essential minerals from the brain and the trap become permeant in most cases. And facilitates other illnesses beside the mental one.

An update on this situation: On June 25, 2019 PBS documentary on the use of methamphetamines to create a super solder. During the Battle of the Bulge the troops were supplied 35,000,000 pills and injections before going into battle.

The British heard of this and to not be out drugged the British high command ordered 20,000 pills for the troops fighting at in Africa. Before a battle with Rommel they were given 20 milligrams of meth. The pills were instant release and they went into battle wired to the max. One of the effects of this much meth was the suspension of the reality of the situation and soldiers felt themselves brave and bullet proof. 89 percent of these troops were killed.

The end of the war did not stop the drug invasion of America: The long-term loser of war if you understand that the corporatism and globalism of the bankers, the objective of the war, was to create a war economy and America was the resource rich and very drug able people were primed for the world police invasion. Large scale protectionism for a fee, like it or not

The invasion plan went well. The scientist of the 'Paper Clip' invaders were soon directing the labs that became the pharmaceutical invaders sent to weaken the will of the people and enslave as we were warned by presidents. First the planting of future disease potential with the atomic bombs was the first step

The German invasion of America began before WW1 with the arrival of bankers and a little later in the 1920s, a media mind manipulator a man called Bernay, Sigmund Frauds' nephew.

Fraud was one of the mind control planners using his understand of the human emotion and program he knew there was a way through drug to control masses. He was involved in the plan of the takeover of America for the controllers use to take over the world for them. America is stupid by plan.

I write of Bernay in this book. He was the man sent to infiltrate the Public Health Service and began the fluoride dumbing of as many as possible with fluoride, a mind control drug.

The DuPont family was the one financing this invasion of German operatives, Albert Einstein was the most well-known invader in scientism clothing.

Edward Teller another. They both, and many others, came disguised as refugees but were part of the plan to trick America to fund its own destruction and mind control.

They were sent to infiltrate the government and in fact persuaded Roosevelt to fund 100 million dollars to build the contamination program with atomic bombs of specific isotopes with known disease causing assurance.

This was brilliant. Send 'fleeing German-Jewish scientists' telling a woe is me story and simultaneously take over the media and are sold as good guys.

Essential to this long-term plan was the establishing a controlled media to sell them to the public. They were German invaders. They were fifth columnist, the hidden secret enemy.

Eisenhower warned that the 'sci-techno elite' were the greatest enemy of the people. The Sacker family were operatives of this drug invasion. The controlling effects of the drugs on people were there objective. This family are operatives of the secret plan to enslave the world. It is working well.

All major meth makers in Mexico and other places get their base chemicals from companies funded to invade with chemistry.

1. Lindner, Kurt
2. Jungert, Wilhelm
3. Debus, Dr. Kurt
4. Fischel, Dr. Edward
5. Gruene, Dr. Hans F.
6. Mrazek, Dr. William
7.
8. Schlitt, Dr. Helmuth
9. Axter, Dr. Herbert
10. Vowe, Theodor K.
11. Beichel, Rudolf
12. Helm, Bruno K.
13. Holderer, Oscar
14. Minning, Rudolf
15. Friedrich, Dr. Hans
16. Haukohl, Guenther H.
17. Dhom, Friedrich
18. Tessmann, Bernhard
19. Heimburg, Karl L.
20. Geissler, Dr. Ernst
21. Duerr, Friedrich
22.
23. Milde, Hans W.
24. Luehrsen, Hannes
25. Patt, Kurt E.
26. Eisenhardt, Otto K.
27. Tschinkel, Dr. J.G.
28. Drawe, Gerhard P.
29. Heller, Gerhard
30. Boehm, Josef
31. Muehlner, Dr. J.W.
32. Rudolph, Dr. Arthur
33. Angele, Wilhelm
34. Ball, Erich K.
35. Heusinger, Bruno K.
36. Novak, Max E.
37. Mueller, Dr. Fritz
38. Finzel, Alfred J.
39. Fuhrmann, Herbert
40. Stuhlinger, Dr. Ernst
41. Guendel, Herbert
42. Fichtner, Hans
43. Hager, Dr. Karl
44. Kuers, Werner R.
45. Bergeler, Herbert
46. Maus, Hans H.
47. Schwidetzky, Dr. W.
48. Hoelker, Dr. Rudolf
49. Kaschig, Erich K.
50. Rosinski, Werner
51. Scharnowski, Heinz
52. Vandersee, Fritz
53. Urbanski, Arthur
54. Tiller, Werner
55. Woerdemann, Hugo
56. Schilling, Dr. Martin
57. Schuler, Albert E.
58. Lindenmayr, Hans J.
59. Zoike, Helmut
60. Paul, Hans G.
61. Rothe, Heinrich C.
62. Roth, Ludwig
63. Steinhoff, Dr. Ernst
64. Reisig, Gerhard H.
65. Klauss, Ernst K.
66. Weidner, Dr. Hermann
67. Lange, Hermann
68. Paetz, Robert
69. Merk, Helmut
70. Jacobi, Walter W.
71. Grau, Dieter E.
72. Schwarz, Friedrich
73. Von Braun, Dr. Wernher
74. Wittmann, Albin E.
75. Hoberg, Otto A.
76. Schulze, William A.
77.
78. Thiel, Dr. Adolf K.
79. Wiesman, Walter
80. Buchhold, Dr. Theodor
81. Rees, Dr. Eberhard H.
82. Hirschler, Otto
83. Poppel, Theodor A.
84. Kroll, Gustav A.
85. Voss, Werner E.
86. Beier, Anton
87. Zeiler, Albert
88. Schlidt, Rudolf H.
89. Steurer, Dr. Wolfgang
90. deBeek, Gerd W.
91. Millinger, Heinz
92. Dannenberg, Konrad K.
93. Palaoro, Hans R.
94. Neubert, Erich W.
95. Sieber, Dr. Werner
96. Hellebrand, Emil A.H.
97. Hosenthien, Hans H.
98. Bauschinger, Oscar
99. Michel, Dr. Joseph
100. Scheufelen, Claus
101. Burose, Walter
102. Fleischer, Karl
103. Gengelbach, Werner
104. Beduerftig, Hermann M.
105. Hintze, Guenther

Here are the "SCIENTIFIC-TECHNOLOGY elite" walk in invaders who took control of America's system to change the mind and enslave us chemically. Few other countries allowed this chemical invasion easily.

"Better living through chemistry" is an old hippie saying but our meaning then was quite different. Brilliant. But lethal. These controllers are like gambling addicts. They slimed into control of the worlds' life support system and they are using air and water pharmacology chemistry to control us. Brilliant but evil because they are using universal principles of *"cause and effect"* by chemical manipulation and another adage is "everything vibrates". Our emotions are like notes of music manipulate our vibrations with cell signals that correspondent to emotions and sometimes actions can be triggered by a signal of vibratory energy.

Just like a note of a song can trigger dance.

The right song note can make killing seem natural.

The known patient potential of iodine bombs was known and a very popular surgery today, and stupid, is the removal of the thyroid gland and the financial imprisonment of having to take drugs for the rest of your life. As planned.

Cesium-137 deposition density due to global fallout.

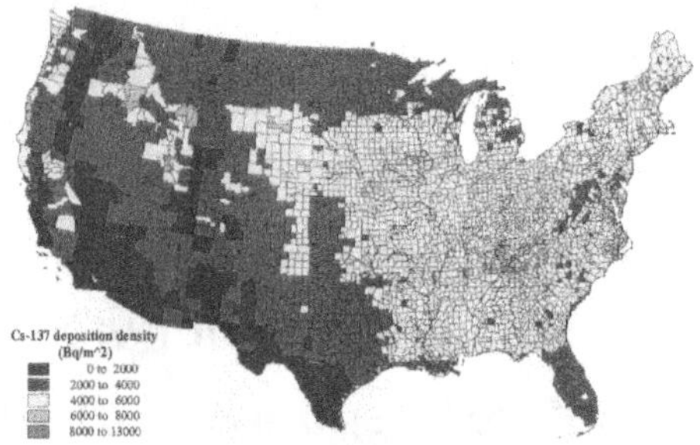

USA Epa statement regarding cesium 137 ingestion. It will cause cancer and shorten-life spans of all humans and mammals. The process in underway in the fall out regions. I was a down winder living in Arkansas where massive fall occurred Christmas 1962

200,000 tons of Iodine debris was flooded into the wind and Arkansas was hit hard. There were 212,000 cases of thyroids disease within months and counting. Those bombs spawned the thyroid medical industry to boom.

HERE WAS THE BEGINNING OF THE THYROID EPIDEMIC AND FAD SURGICAL REMOVAL OF A VITAL GLAND MAKING THE PATIENT A FOREVER SLAVE OF THE BIG PHARMA

Some of the Creeps involved in this conspiracy

VINDICATION OF JFK

CONSPIRACY WARNING SPEECH

CLEARIFICATION OF THE WHO BEHIND THE FINANCING OF THE MILITARY INDUSTRAL COMPLEX MENTIONED BY EISENHOWER IN HIS LAST SPEECH

A master plan by 1000 dream of Hitler planners to control American resources and enslave all chemically to do their bidding is in place. Controllers have installed through the controlled media the mindless ***programmed*** warring response to the key words “fighting for democracy and freedoms”.

Programming all Christian, and some American Jews to killing and taking the freedom from others while stealing their resources.

Kennedy was killed before he could truly expose the “conspiracy” faces behind the planned long term and take of this resource called America.

Main among the devils were the German bankers, the Warburg family and friends including the breeders of future presidents, Democrat and Republican!

They came here as part of the real invasion that took place, through the Warburg’s financial manipulations including investment in booze, beginning with the designing and execution, of the **German** Federal Reserve System.

We are controlled to this day by the system they built. Indians were the first target of the predecessors of these monsters.

When Kennedy blew the whistle, and used

the term **conspiracy**

the **German** Warburg Media controllers went to work discrediting anyone thinking out loud that there was such a conspiracy.

They erased most records of this in the Library of Congress and across the world 's mind. Yeah, so cool such power!

To this day I still hear the program coming out of many when I say there is a conspiracy.

I am immediately verbally assaulted with ridicule. But here is the evidence. I kept looking in old archives and here is the evidence that conspiracy trials of 1935 were held and a "redacting" (Notice the term) congressional committee hid the evidence.

And those who financed the coup attempt continued. They were the financial wave of the true invasion still breaking over the American conscience covering at high tides this reality of a dominion plan that includes you.

Drop the Jew shield that is automatic in those hiding the truth.

These are Germans Aryan Supremes and they brought Germans hiding behind holocaust sympathy to take over our system of government and it is working splendid.

Culture trace in action to fool the stupid gentiles and the gullible everyone else.

When this "conspiracy situation became a "national furor in 1935" they went underground and continued to manipulate.

Over time those who remember this situation were dead or dying by the time Kennedy tried to tell us a profound truth.

Here is the cool part of this article. There is clear expose' of these wealthy German/Jewish bankers.

They financed the work camps where 1 out of 50 worked to death were Jews.

So why the ongoing woe is me program that we are hit with at every turn. Goddamn, we see the twisted and probably enhanced Anne Frank story that is told over and over and never a film about the other humans who were worked to death. The program in motion aka pictures.

That is programing and in another generation, the German cover, Jewish version, story of the holocaust will be totally programed into all education and media.

The President Eisenhower speech telling us that no Jewish holocaust happened is buried and is not taught like the myth of a Jewish only holocaust.

In time, it will seem to history readers that the Super German Jews or “Royal Jews” were the only real people who own Earth and must be obeyed rulers, just like the “subjects” in England, gentiles and others.

The English of Germanic origin are still not our real friends. They came in 1607, made profits from wars and drugs, learned the language so to speak, and stayed.

Remember BP Horizon. They also have ‘chemo-planes’ bases and not chemo-crop-dusters. “Cyprus skies are filled often with

British Airforce planes spraying chemo" into the air and rain.

Brilliant mind control. I am tired of it.

This books' Jewish references is not about the Jewish man and family who works and lives and loves.

It is about the German invaders and shows that Jews, Gentiles and all other breathing the air are part of this medical cartel business project of profit.

Whether you like it or not. Hitler a part Hebrew DNA/German had a 1000 years dream of Universal Master over all minds on this planet.

NASA is aggressively and fatal to the planet has planes to invade all universe.

Hitler's Dream in motion carried by Von Braun still invades the 'Bank of American peoples' accounts for a foolish dream, lethal to all players not given anti-dotes to the planned toxicity culling of 5 billion

ASAP. They hide within all religions and Judaism has big skirts to hide within and use to shield their controller madness.

"We shall propose further cooperative efforts between all nations in weather prediction and eventually in weather control."
- President John F. Kennedy
Sept. 25, 1961
WEATHER
WARFARE

Yeah pic twice, he was good a ghost mentor I do not hold this mistake against him.

Kennedy is Vindicated

I saw him in Little Rock and saw his speech.

Do not remember words just the man

A German Scientist and Nazi with JFK

This photo was taken days before he was killed for attempting to expose this man's part in a "Conspiracy' to rule the Earth

You think maybe Kennedy ask why and did not like the answers. They killed him when he went public.

There is hope in Texas. At the Plaza where "they" killed him is a Kennedy museum that now has exhibits the true story. There were several gunmen and the Secret Service was involved in the set-up.

A

Conspiracy Exposed

Truly

WARNING FROM PRESIDENT EISENHOWER, A JEW, THE GIST, DO NOT TRUST THE POWER MAD GERMAN SCIENTISTS

Eisenhower

—"world peace and human betterment."

"Yet, in holding scientific research and discovery in respect, as we should, we must also be alert to the equal and opposite danger that public policy could itself become the captive of a scientific-technology elite"

"We want democracy to survive for all generations to come, not to become the insolvent phantom of tomorrow

KENNEDY

"If you are awaiting a finding of "clear and present danger," then I can only say that the danger has never been more clear and its presence has never been more imminent

It requires a change in outlook, a change in tactics, a change in missions--by the government, by the people, by every businessman or labor leader, and by every newspaper".

"For we are opposed around the world by a monolithic and ruthless conspiracy that relies primarily on covert means for expanding its sphere of influence–on infiltration instead of invasion, on subversion instead of elections, on intimidation instead of free choice, on guerrillas by night instead of armies by day."

“It is a system, which has conscripted vast human and material resources into the building of a tightly knit, highly efficient machine that combines military, diplomatic, intelligence, economic, scientific and political operations”

Its preparations are concealed, not published. Its mistakes are buried, not headlined. Its dissenters are silenced, not praised. No expenditure is questioned, no rumor is printed, and no secret is revealed. It conducts the Cold War, in shorts; with a wartime discipline, no democracy would ever hope or wish to match”.

A CONSPIRACY?

Can our beloved past Presidents?

Be among those who knew and understood

That an invasive organized plan was in motion

Around the world

One capable of covering its track

While operating in the open

Could it be true after all that a “conspiracy”?

Or an organized business plan

Of a fortunate but mentally corrupted few

Designed to control the people of the world

Be unfolding every hour?

What was it they were warning us about?

Preparing us to confront

What hidden in plain sight force

Was a foot in America?

I now know the truth that Hitler’s 1000 Dream of World Dominion is almost revealed

This world dominion business plan, often hiding behind innocent Hebrews.

Adolph Killed THE Kennedys

WALL STREET'S FASCIST CONSPIRACY

Testimony that the Dickstein Committee Suppressed

JOHN L. SPIVAK

AN ORGANIZED conspiracy exists to seize the government by a fascist coup. The Congressional Committee appointed to investigate just such activities has not only failed to follow the trail of evidence to its fountain head—Wall Street—but has deliberately suppressed evidence pointing in that direction.

In these articles the reality of Wall Street's fascist conspiracy will be made clear; the line-up of financial interests back of the conspiracy will be set forth; and the real role of the Dickstein Committee, which suppressed this evidence, will be revealed.

A suggestion of the existence of Wall Street's fascist conspiracy was made public in November. The Dickstein Committee then was forced to call Gen. Smedley D. Butler, one of those who made the charges, to testify. And that was the end of the Committee's interest in proving the charges.

This series of articles will go deeply into the whole situation, of which only a hint trickled through to the public. The suppression of evidence by the Dickstein Committee reveals the Committee's real character: With an ostensible mission to uncover fascist activities, the Committee actually turned out to be a close collaborator with the would-be fascist rulers of the country; it covered up the conspiracy by suppressing evidence which led too high up in those financial and industrial groups which run Congress, "advise" the President, and dominate the country.

It will be shown that financial and economic class considerations rise above every other kind, including racial and religious ones. The anti-semitic character of Nazism has been abundantly demonstrated in these pages; nevertheless this article, and succeeding ones, will reveal Jewish financiers working with fascist groups which, if successful, would unquestionably heighten the wave of Hate-the-Jew propaganda.

The class basis of social forces is nowhere more clearly revealed than in this situation—capitalists, including Jews, making common cause with anti-semitic fascist and potentially fascist organizations, in an effort to crush labor. The ultimate aim, of course, is the true fascist one of a "totalitarian state," with all cultural, educational, and political activities inimical to capitalism suppressed. The immediate path to this objective is the destruction of the labor movement and particularly the militant vanguard represented by the Communist Party.

The Dickstein Committee has deliberately suppressed testimony of fascist activities which it had in its possession. This evidence was suppressed because financial powers behind the committee are among the supporters of fascist organizations.

Throughout its investigation of Nazi, fascist and Communist activities the Committee has been careful not to involve certain financial interests—such as J. P. Morgan and Co., Kuhn, Loeb and Co., etc. Felix Warburg, head of the Kuhn, Loeb banking house virtually dominates it as well as the American Jewish Committee, a powerful organization active in fighting the spread of anti-semitism. The American Jewish Committee is controlled by wealthy Jews. However, it has a large following among lower and middle-class Jews who are not aware of the maneuverings of the leadership for its own economic interests. The zeal of the leaders in fighting anti-semitism is tempered by the financial interests of some of them—in the United States and in Nazi Germany—and by the active participation of some of them in fascist organizations in this country.

Shortly after the Dickstein Committee was empowered by Congress to investigate "subversive" activities, leaders of the American Jewish Committee began to steer the Congressional Committee's investigations. In the course of this steering, information was suppressed which reflected upon leading bankers, as well as information of fascist organizations in which they were interested.

Instead of actually seeking evidence of fascist organizations and who are behind them, the Congressional Committee ignored Fascism until its menace here was thrust upon them; and then suppressed vital evidence regarding it. The reason: Wall Street interests such as Morgan's were involved which are tied up with the Warburg interests—which dominate the American Jewish Committee without the knowledge of the overwhelming majority of its membership.

In the course of these articles I shall show:

1. That the Dickstein Committee refuses to explain why it suppressed evidence of fascist organizations and of fascist movements.

2. That the Dickstein Committee knew of the offer made to Gen. Smedley Butler to organize a fascist army of 500,000 men, but ignored this information until it was forced to call Butler.

3. That having called him, the Committee issued a garbled statement of what he said and not until the national furore died down did it issue even *parts* of his testimony.

4. That Gen. Butler named a fascist organization in which some leaders of the American Jewish Committee are active—and that this testimony was suppressed.

5. That a Nazi agent worked in Warburg's Bank of Manhattan and that Felix Warburg was never called upon to explain how he got there.

6. That the Warburg financial interests have heavy investments in Nazi Germany. The American Jewish Committee has steadfastly opposed the boycott of German goods.

7. That the most powerful fascist organizations are controlled by financiers whose interests are controlled by J. P. Morgan's interests.

8. That the Warburg financial interests are tied up with Morgan and consequently work with Morgan men.

9. That Grayson M-P Murphy, involved in the plot to organize a *fascist army*, is a Morgan man and one of those who originally financed the starting of the American Legion for "Big Business" and who supports dissemination of anti-semitic propaganda; and that, knowing all this the Dickstein Committee never called Murphy to explain his activities.

10. That a Hearst man tied up with Morgan interests captured control of the American Legion, which Butler was asked to lead as a fascist army; and that this man, summoned to appear before the Dickstein Committee, was never questioned after he had had a secret conference with President Roosevelt.

11. That the American Liberty League was named by Butler and this fact suppressed by the Dickstein Committee. The League is controlled by Morgan-du Pont interests as well as having Warburg representation on it.

12. That the Remington Arms Co., controlled by Morgan-du Pont, was named as the body which would supply arms and equip-

ment to the fascist army and that this testimony was suppressed by the Congressional Committee.

13. That Max Warburg, brother of Felix, and director of the steel trust of Germany, which originally financed Hitler, are in the United States trying to get credits for Hitler's government in copper purchases.

14. That Hearst copper interests were among those being considered at the time Hearst opened his anti-red campaign.

Let us first consider Butler's testimony that he was offered $3,000,000 to organize a fascist army with a promise of $300,000,000 more if it became necessary. I shall review it very briefly to refresh the reader's mind.

Gen. Smedley Butler testified that he was approached by Gerald C. MacGuire, a "$100 a week bond salesman," with an offer of $18,000 in one thousand dollar bills to go to the American Legion convention in Chicago in 1933 to make a speech in favor of the gold standard; it was after this connection was established that MacGuire suggested organizing the fascist army. MacGuire at that time said he was working for Robert S. Clark, who inherited millions of the Singer Sewing Machine fortune. While working for Clark, MacGuire was kept on the payroll of Grayson M-P Murphy, a "Wall Street broker." During the period when these negotiations were going on, MacGuire, who had never owned more than a few thousand dollars, suddenly began to handle large sums of money, depositing and withdrawing amounts running or beyond $100,000. The Dickstein Committee, in examining him, found that he could not account for $65,000 which were spent during the trip to the American Legion convention and that he lied repeatedly about what he had done with certain large sums.

So much for the Butler story; what is not known is that long before General Butler testified, the Congressional Committee investigators knew about it. Nevertheless they did not call Butler, though one of the things they were supposed to investigate was "subversive" activities, including Fascism. The Dickstein Committee called Butler only when it learned that The New York Post and the Philadelphia Record were about to publish the story anyway, which they had learned through their reporter Paul Comley French, a friend of the General's.

The national furore aroused by the story was so great that the Committee had to issue a statement after getting the testimony in secret session. When the excitement died own the Congressional Committee issued a mmary of the Butler testimony for the ress, Butler having been cautioned not to divulge what happened behind the committee's closed doors, according to the General.

During the course of my investigation into fascist activities in the United States, I persistently asked for the Butler testimony. I told that "the summation tells the whole ory."

"But why can't I see the whole testimony? What is there in it which you do not want me to see?"

General Smedley Butler quoting Robert S. Clark, who sent Gerald C. MacGuire with proposals for a fascist army (the suppressed testimony is in italics):

The Published Testimony:	***What Butler Really Said:***
He (Roosevelt) has either got to get more money out of us or he has got to change the method of financing the Government and we are going to see to it that he does not change that method. He will not change it.	He (Roosevelt) has either got to get more money out of us, or has got to change the method of financing the Government, and we are going to see to it that he does not change the methods. He will not change it. *He is with us now."*
I said, "The idea of this group of soldiers then, is to sort of frighten him, is it?"	I said, "The idea of this great group of soldiers, then, is to sort of frighten him, is it?"
"No, no, no, not to frighten him. This is to sustain him when others assault him."	"No, no, no; not to frighten him. This is to sustain him when others assault him."
	He said, "You know, the President is weak. He will come right along with us. He was born in this class. He was raised in this class, and he will come back. He will run true to form. In the end he will come around. But we have got to be prepared to sustain him when he does."
I said, "Well, I do not know about that. How would the President explain it?"	I said, "Well, I do not know about that. How would the President explain it?"

"Nothing has been left out, except some hearsay evidence," I was assured. "A few names were mentioned which have nothing to do with the case."

After my persistence had made it clear that my suspicions were growing, I was handed a copy of the hastily published Butler testimony, marked "extracts." At the end of the 125 page record was a note in bold face type:

> **The Chairman: In making public the foregoing evidence, which was taken in executive session in New York City November 20 to 24, inclusive, the Committee has ordered stricken therefrom certain immaterial and incompetent evidence, or evidence which was not pertinent to the inquiry, and which would not have been received during a public hearing.**

The printed question-and-answer testimony gave more information than the summation originally issued by the Committee. I was still curious to know just what "evidence" the Committee considered "immaterial," my curiosity being heightened when I was told by a person in a position to know and who had never told me anything unfounded, that the request to suppress certain parts of Butler's testimony had come from Henry Morgenthau, Secretary of the Treasury. I could not prove it but I had enough faith in my informant to believe it.

More requests for the uncensored stenographic notes of Butler's testimony met with refusals. The "immaterial evidence" was a carefully guarded secret. Eventually I did obtain these suppressed stenographic notes. With the notes in my possession as well as knowledge of the financial interests within and behind the American Jewish Committee, the leaders of which were steering the Congressional Committee, I called upon the chairman, Congressman John W. McCormack. I had prepared a series of questions for the interview which he had agreed to give me. When I got to the sixth question which probed a little deeper into the suppression of evidence by his Committee, the Congressman became a little nervous.

"Oh, somebody's been telling you things," he said.

"No, no one has been telling me things. I have the stenographic notes."

"Those are executive minutes," he exclaimed. "I can't imagine how they got in your possession. I must find out."

The knowledge that I had the suppressed testimony obviously upset him. The interview had been progressing in a friendly manner until I got to the stage where it seemed that a Congressional investigating committee was being investigated. Suddenly he said abruptly:

"I don't have to answer your questions."

"That's right," I assured him, "you don't."

"And I don't have to give you an interview."

"That's right, too."

"Well then, cancel this interview."

"Okay, I'll cancel it. But don't you think you had better answer the questions?"

"I will not answer any more questions. It is obvious to me that they are cleverly arranged—all leading to one point—you want to hang me."

"No, I don't want to hang you. I think your committee has hanged itself."

"I'll take your notes and the questions and answer such of them as I wish. I want to think them over."

"That's okay," I agreed, handing him the questions.

Some of the brief questions I asked him follow:

Will you define what you mean by Nazism, Fascism, Communism?

Did you ever look into the potential fascist groups like the American Liberty League, Father Coughlin's organization, the Crusaders, etc.?

Did you ever investigate why the American Legion passed the gold resolution while MacGuire was in Chicago with a lot of money?

Why wasn't John Taylor called regarding Legion and Veterans of Foreign Wars activities?

Why didn't you investigate the educators' charges that Hearst was carrying on fascist propaganda?

What relationship has your Committee with the American Jewish Committee?

When The New Masses published evidence that Ralph Easley of the National Civic Federation was secretly reporting to George Sylvester Viereck, the Nazi agent, while the former was distributing Communism in Germany, why weren't Easley's finances looked into?

The Jewish-controlled concerns in Chicago who contributed to Harry A. Jung's organization and the money used to disseminate anti-semitic propaganda were known to you. Why wasn't that evidence made public?

Why wasn't the relationship between Kuhn, Loeb and Max Warburg established to determine why a Nazi agent found his way into Warburg's Bank of Manhattan? Why wasn't Felix Warburg questioned about it?

Did you ever investigate the financial dumps of bankers and industrialists to determine the motivation in supporting potential fascist groups?

Did you ever investigate Assistant Secretary of War Woodring's statement that the C.C.C. boys would be "economic storm troops" against "social disorders?"

Did you ever investigate why organizations which started out for monetary reform like the Committee for the Nation ended up by carrying on anti-labor propaganda?

Did you ever question Under-Secretary of State Phillips why he met with Easley to try to stop the boycott of German goods and thus give economic aid to the Nazis?

Did you ever get to the bottom of the report that John W. Davis wrote the gold speech passed at the Chicago convention?

I agreed not to use the statements he had made before he cancelled the interview and I gave him the questions with my notes on them. He promised to give me written answers to "those he wanted to answer" within three days. On the day he promised his answer I got it. I read it over five or six times. I still don't know what he is talking about. For the reader's benefit I give his answers. I think they show the state the chairman is in:

My dear Mr. Spivak:

On Saturday last you called into my office for an interview, as a result of which you left with me a number of questions which you intended to ask me. I told you that I would consider them and write you on or before the following Tuesday. I am complying with what I told you, to write you on or before Tuesday Jan. 15.

Pending the report of the committee to the use of Representatives I have discussed only limited way with representatives of the press opinion of the value of some evidence obtained, and the probable recommendations that impressed me personally. The final report and recommendations will be determined later by the full committee.

Assuming the premises upon which they are predicated are correct, and it is plain to see that they are not, some of the questions that you intended to ask relate to matters beyond the jurisdiction of the committee, and its powers of investigation. I, therefore, ignore them.

There are some questions which you intended to ask which I would have no hesitancy in answering if asked by other representatives of the press, and while I am not in sympathy with the policies, associations or affiliations of the publication which you represent, naturally, I would accord you the courtesy and consideration I would extend to others.

You were particularly anxious to find out if the Nazi movement in this country is as active today as it was when the investigation started. As a result of the investigation, and the disclosures made, this movement has been stopped, and is practically broken up. There is no question but what some of the leaders are attempting to carry on, but they can make no headway. Public opinion, as a result of the disclosures of the investigation is aroused.

The breaking-up of any intolerant movement, the objective of which is to group Americans against Americans, or persons against persons, because of race, color or creed, is beneficial to the country and the people as a whole. The same opinion applies to a movement dedicated to the overthrow of government by legal or illegal means, or a combination of both, employing force and violence, if necessary to obtain the desired objective. The use of lawful or legal means is a right which every person or movement possesses to change, in whole, or in part, our government, even though one may not agree with the methods employed, or the purposes and objectives of such a movement. No person or movement has a right to resort to illegal means to accomplish this end. When such methods are employed, the resort to violence and force, to try and obtain the overthrow of government, whether or not it is or can be accomplished, it is beyond the pale of the Constitution, and of rights guaranteed thereunder.

The reason for certain portions of General Butler's testimony in executive session being deleted from the public record has been clearly stated in the printed public record.

Very truly yours,

John W. McCormack.

All I can say regarding this is that I hope the Committee's report to the Congress will be clearer.

Still searching for the Committee's explanation of why it suppressed testimony of Fascism and fascist organizations, I called upon Congressman Samuel Dickstein, vice-chairman of the Committee on "un-American" activities. Like many others, I refer to this Congressional body as the "Dickstein Committee," chiefly because Dickstein first introduced the bill for the investigation; but calling it the "Dickstein Committee" is a misnomer...

Gen. Smedley Butler quoting MacGuire, who, the General testified, came to him with an offer to lead a fascist army (the suppressed testimony is in italics):

The Published Testimony:

I said, "Is there anything stirring about it yet."

"Yes," he says; "you watch; in two or three weeks you will see it come out in the paper. There will be big fellows in it. This is to be the background of it. These are to be the villagers in the opera. The papers will come out with it." He did not give me the name of it, but he said it would all be made public; a society to maintain the Constitution, and so forth. They had a lot of talk this time about maintaining the Constitution. I said, "I do not see that the Constitution is in any danger," and I asked him again, "Why are you doing this thing?"

We might have an assistant President, somebody to take the blame; and if things do not work out, he can drop him.

What Butler Really Said:

I said, "Is there anything stirring about it yet."

"Yes," he says; "you watch; in two or three weeks you will see it come out in the paper. There will be big fellows in it. This is to be the background of it. These are to be the villagers in the opera. The papers will come out with it," *and in about two weeks the American Liberty League appeared, which was just about what he described it to be. That is the reason I tied it up with this other thing about Al Smith and some of these other people, because of the name that appeared in connection with this Liberty League.* He did not give me the name of it, but he said that it would all be made public. . . .

We might have an assistant President, somebody to take the blame; and if things do not work out, he can drop him. *He said, "That is what he was building up Hugh Johnson for. Hugh Johnson talked too damn much and got him into a hole, and he is going to fire him in the next three or four weeks."*

I said, "How do you know all this?"

"Oh," he said, "we are in with him all the time. We know what is going to happen."

Paul Comley French, reporter for the New York Post, telling of his conversations with Gerald MacGuire (the suppressed testimony is in italics):

The Published Testimony:	What French Really Said:
At first he (MacGuire) suggested that the General organize this outfit himself and ask a dollar a year dues from everybody. We discussed that, and then he came around to the point of getting outside financial funds, and he said that it would not be any trouble to raise a million dollars.	At first he (MacGuire) suggested that the General organize this outfit himself and ask a dollar a year dues from everybody. We discussed that, and then he came around to the point of getting outside financial funds, and he said that it would not be any trouble to raise a million dollars. *He said that he could go to John W. Davis or Perkins of the National City Bank, and any number of persons and get it.*
	Of course, that may or may not mean anything. That is, his reference to John W. Davis and Perkins of the National City Bank.
	During my conversation with him I did not of course, commit the General to anything. I was just feeling him along. Later we discussed the question of arms and equipment, and he suggested that they could be obtained from the Remington Arms Co., on credit through the du Ponts. I do not think at that time he mentioned the connections of du Pont with the American Liberty League, but he skirted all around it. That is, I do not think he mentioned the Liberty League, but he skirted all around the idea that that was the back door, and that this was the front door; one of the du Ponts is on the board of directors of the American Liberty League and they own a controlling interest in the Remington Arms Co. In other words he suggested that Roosevelt would be in sympathy with us and proposed the idea that Butler would be named as the head of the C.C.C. camps by the President as a means of building up this organization. He would then have 300,000 men. Then he said that if that did not work the General would not have any trouble enlisting 500,000 men.

have injustice to the Congressmen. It is not his committee. The financial powers in the American Jewish Committee, which directed Congressional body, simply played circles and the bewildered Congressman. Dickstein never knew, and I doubt if he knows now, just what happened and why certain specific evidence was suppressed. Throughout the whole investigation he kept blundering into things which shouldn't have been blundered into and he could never understand why those steering the Committee opposed probing along lines which would lead to the Warburg-Morgan interests. When I talked with him and pointed out the financial hook-ups he looked sad.

"I wish you had told me that while the Committee was in session," he said plaintively. "I'd have called Murphy and Morgan and Warburg and anyone else involved."

Dickstein's activities in the Committee—such questions as he persisted in asking—were chiefly confined to the Nazis. Communism was really dragged into this investigation; and the financial powers behind the Congressional Committee certainly had no intention of investigating Wall Street's fascist conspiracy until the threat of breaking the Butler story in the press forced them to make a gesture in that direction. The investigation into Communism was steered by the leaders of the American Jewish Committee, Felix Warburg and his non-Jewish Wall Street colleagues, for three reasons:

The growing interest in and sympathy the Communist movement in industries where these financial powers had investments; if the Communist Party could be outlawed it would be of tremendous advantage to the financiers and industrialists guiding the work of the Committee.

2. There was a great deal of publicity in the press and propaganda by Nazi agents that "a Communist is a Jew and a Jew is a Communist."

3. A federal law ostensibly directed at Communists as "subversive elements" could be used to keep labor from doing a great many things, whether labor was affiliated with left-wing organizations or conservative ones like the American Federation of Labor.

Even William Green, president of the A. F. of L., realized that. I was present at the hearing in Washington when Green testified and it was really one of the funniest shows I ever saw. There is nothing that Green would like better than to see the Communist Party outlawed, but Green realized that any such procedure would be directed at all labor, and would eventually endanger his own position. *Dickstein* and McCormack, neither having a fraction of the knowledge of the labor movement that Green has, tried in a dozen different ways to get Green to say that a bill outlawing the Communist Party would be a good thing—and Green persistently assured that any such move would react, against ..F. of L. and would be fought.

It was a very depressing hearing for both ..-. and Dickstein (who had competed with one another for the most publicity during the life of the investigation). The two Congressmen had issued statements that they intended to outlaw the Communist Party months before the Committee had finished its investigation!

A good idea of the stature of Dickstein can be had by his answers to some of the questions I asked him.

"Congressman, just what do you mean by Nazism?" I asked.

"Well, Nazism is—you see—you know I'd rather you'd get the definition I gave of it in my last speech."

"Okay. How about Fascism."

"That's in there, too."

I tried again.

"Do you think Fascism is the last stand of capitalism?"

"Certainly," he said. "Powerful wealth is concentrating for its own preservation."

"And your committee was supposed to investigate Fascism?"

"Yes, Fascism. All subversive, un-American movements."

"A real investigation of Fascism or fascist movements in this country would have to take in a study of powerful financial groups and their motivations?"

He looked at me warily, as though fearful of a trap, and nodded solemnly.

"Then why didn't the Committee investigate the financial tie-ups to determine the motives behind such groups as the American Liberty League?"

"Well, we didn't have the time or the money, or we would have."

"What was left out of the Butler story

"We confined our activities to evidence permissible in a court. We didn't go into the details because it was hearsay."

"But your published records are full of hearsay evidence."

He looked at me, startled.

"They are?"

"Well, why wasn't Grayson M-P Murphy called? Your committee knew that Murphy's men are in the anti-semitic espionage Order of '76; it knew that Murphy was supporting Edmondson in sending out his anti-semitic news releases; it knew that Murphy and Clark were hooked up for years selling bonds together—why wasn't Murphy called?"

"We didn't have the time. We'd have taken care of the whole Wall Street group if we had had the time. I would have no hesitation in going after the Morgans."

"Did you ever go into the fascist or potentially fascist—groups like the American Liberty League, the Crusaders, etc?"

"No, we went a little into the Black Shirts—it's an organization like the Nazis but it didn't amount to anything. We had no time," he repeated.

"You had Frank Belgrano, commander of the American Legion, listed for testimony. Why wasn't he examined?"

"I don't know," he said. "Maybe you can get Mr. McCormack to explain that. I had nothing to do with it."

"Why didn't you call Easley after THE NEW MASSES had published his secret reports to George Sylvester Viereck, the Nazi agent, and find out about Easley's finances?"

"To the best of my recollection, Easley was called into executive session. He testified about Communism."

"I don't doubt it. But I'm interested in why his finances were not examined since he was distributing an anti-semitic book imported into this country by Viereck."

"I don't know."

"Why weren't the names of the Jewish concerns whose money went to Harry A. Jung in Chicago and which was used for anti-semitic propaganda, made public?"

"I never saw them," he said. "We have so much stuff I haven't had a chance to read all the reports. I wasn't at the Chicago hearing."

"And McCormack wasn't at the Chicago hearing. Then who issued orders not to make those names public?"

"I don't know."

"Why wasn't Edward A. Rumely questioned regarding the Committee for the Nation activities which benefited Nazi Germany and on whose committee Lessing Rosenwald of the American Jewish Committee was active?"

"I couldn't answer that. You'd have to ask McCormack about it."

"Okay. Why wasn't Felix Warburg questioned as to how the Nazi agent F. X. Mittmeier got a job in the Warburg-controlled Bank of Manhattan?"

"I don't know."

"Fascism came at the last moment," he said, switching the subject. "I knew of only one fascist group—the Black Shirts—and they weren't important."

"Didn't Assistant Secretary of War Woodring's statement that the C.C.C. boys would be used as 'economic storm troops against social disorders' sound like Hitler Fascism? Why wasn't Woodring questioned about it?"

"There was no time," the Congr. said dazedly.

"But Woodring is in Washington. As were you."

"Maybe the Committee felt there was evidence—maybe."

It was obvious to me that Dickstein sim did not know what was going on around hi when I pointed out the financial tie-ups of t Warburg interests with Morgan inter which Murphy represents and the Warb group with the American Jewish Commit leadership which was steering the Congr sional Committee he was utterly dazed. The tie-ups will be explained in detail in the nex article.

In the meantime I offer the suppressed te timony.

The Congressional Committee had Gen. Butler behind closed doors in a secret sessio It did not know what Butler might say so it wanted to be in a position to suppress tes timony given under oath if this proved neces sary. And it was, for Butler named persons whom the Committee should have called to check various angles—persons high in the po litical and financial world. There is no nee of my repeating much of the General's testi mony. I shall offer only what the publishe report by the Congressional Committee sa he said and what the carefully guarded ster graphic notes show he really said.

Gen. Butler was telling the story of Mu phy's man, (MacGuire's) talk with hi In the left column is what the Commi. published. In the right column is what h actually said—the suppressed testimony be printed in italics.

The Published Testimony:

Then MacGuire said that he was the chairman of the distinguished-guest committee of the American Legion, on Louis Johnson's staff; that Louis Johnson had, at MacGuire's suggestion, put my name down to be invited as a distinguished guest of the Chicago convention.

I thought I smelled a rat, right away—that they were trying to get me mad—to get my goat. I said nothing.

"He (Murphy) is on our side, though. He wants to see the soldiers cared for."

Well, that was the end of that conversation.

What Butler Really Said:

Then MacGuire said that he was the chairman of the distinguished-guest committee of the American Legion, on Louis Johnson's staff; that Louis Johnson had, at MacGuire's suggestion, put my name down to be invited as a distinguished guest of the Chicago convention; *that Johnson had then taken this list, presented by MacGuire, of distinguished guests, to the White House for approval; that Louis Howe, one of the secretaries to the President, had crossed my name off and said that I was not to be invited—that the President would not have it.*

I thought I smelled a rat, right away—that they were trying to get me mad—to get my goat. I said nothing.

"He (Murphy) is on our side, though. He wants to see the soldiers cared for."

"Is he responsible, too, for making the Legion a strike-breaking outfit?"

"No, no. He does not control anything in the Legion now."

I said: "You know very well that it is nothing but a strike-breaking outfit used by capital for that purpose and that is the reason they have all those big club houses and that is the reason I pulled out from it. They have been using these dumb soldiers to break strikes."

He said: "Murphy hasn't anything to do with that. He is a very fine fellow."

I said, "I do not doubt that, but there is some reason for his putting $125,000 into this."

Well, that was the end of that conversation.

He (Clark) laughed and said, "That speech cost a lot of money." Clark told me that it had cost him a lot of money. He thought it was a big joke that these fellows were claiming the authorship of that speech.

I think there was one other visit to the house because he (MacGuire) proposed that I go to Boston to a soldiers' dinner to be given in my honor. He suggested that I go up to Boston to this dinner for the soldiers. He said, "We will have a private car for you on the end of the train. You will make a speech at this dinner and it will be worth a thousand dollars to you."

I said, "I never got a thousand dollars for making a speech."

He said, "You will get it this time."

"Who is going to pay for this dinner and this ride up in the private car?"

"Oh, we will pay for it out of our private funds."

I am not going to Boston. If the soldiers of Massachusetts want to give a dinner and want me to come, I will come. But there is no thousand dollars in it."

So he said, "Well, then, we will think of something else."

Then when he met me in New York he had another idea....

Now, I cannot recall which one of these fellows told me about the rule of succession, about the Secretary of State becoming President when the Vice-President is eliminated. There was something said in one of the conversations that I had, that the President's health was bad, and he might resign, and that Garner did not want it anyhow, and then this super-secretary would take the place of the Secretary of State and in the order of succession would become President. That was the idea. He said that they had this money to spend on it, and he wanted to know again if I would head it, and I said, "No, I was interested in it, but I would not head it."

He (Clark) laughed and said, "That speech cost a lot of money." Clark told me that it had cost him a lot of money. *Now either from what he said then or from what MacGuire had said, I got the impression that the speech had been written by John W. Davis—one or the other of them told me that*—but he thought that it was a big joke that these fellows were claiming the authorship of that speech.

I think there was one other visit to the house because he (MacGuire) proposed that I go to Boston to a soldiers' dinner to be given *by Governor Ely* for the soldiers, *and that I was to go with Al Smith.* He said, "We will have a private car for you on the end of the train *and have your picture taken with Governor Smith.* You will make a speech at this dinner and it will be worth a thousand dollars to you."

I said, "I never got a thousand dollars for making a speech."

He said, "You will get it this time."

"Who is going to pay for this dinner and this ride up in the private car?"

"Oh, we will pay for it out of our funds. *You will have your picture taken with Governor Smith.*"

I said, "I do not want to have my picture taken with Governor Smith. I do not like him."

"Well, then, he can meet you up there."

I said, "No, there is something wrong in this. There is no connection that I have with Al Smith, that we should be riding along together to a soldiers' dinner. He is not for the soldiers' either. I am not going to Boston *to any dinner given by Governor Ely for the soldiers.* If the soldiers of Massachusetts want to give a dinner and want me to come, I will come. But there is no thousand dollars in it."

So he said, "Well, then, we will think of something else."

I said, "What is the idea of Al Smith in this?"

"Well," he said, "Al Smith is getting ready to assault the Administration in his magazine. It will appear in a month or so. He is going to take a shot at the money question. He has definitely broken with the President."

I was interested to note that about a month later he did, and the New Outlook took the shot that he told me a month before they were going to take. Let me say that this fellow has been able to tell me a month or six weeks ahead of time everything that happened. That made him interesting. I wanted to see if he was going to come out right.

So I said at this time, "So I am going to be dragged in as a sort of publicity agent for Al Smith to get him to sell magazines by having our picture taken on the rear platform of a private car, is that the idea?"

"Well, you are to sit next to each other at dinner and you are both going to make speeches. You will speak for the soldiers without assaulting the Administration, because this Administration has cut their throats. Al Smith will make a speech, and they will both be very much alike."

I said, "I am not going. You just erase that out."

Then when he met me in New York he had another idea....

Now, I cannot recall which one of these fellows told me about the rule of succession, about the Secretary of State becoming President when the Vice-President is eliminated. There was something said in one of the conversations that I had *either with MacGuire or with Flagg, whom I met in Indianapolis,* that the President's health was bad, and he might resign, and that Garner did not want it anyhow, and then this super-secretary would take the place of the Secretary of State and in the order of succession would become President. *He made some remark about the President being very thin-skinned and did not like criticism, and it would be very much easier to do it on some-*

He said, "When I was in Paris, my headquarters were Morgan & Hodges (Harjes). We had a meeting over there. I might as well tell you that our group is for you, for the head of this organization. Morgan & Hodges (Harjes) are against you. The Morgan interests say that you cannot be trusted, that you are too radical, and so forth, that you are too much on the side of the little fellow; you cannot be trusted. They do not want you. But our group tells them that you are the only fellow in America who can get the soldiers together. They say, 'Yes, but he will get them together and go the wrong way.' That is what they say if you take charge of them."

So he left me saying, "I am going down to Miami. . . ."

body else. He could say that he was above such routine matters and let the other fellows take care of it and then get rid of him if necessary. That was the idea. He said that they had this money to spend on it, and he wanted to know again if I would head it, and I said, "No, I was interested in it, but I would not head it."

He said, "When I was in Paris, my headquarters were Morgan & Hodges (Harjes). We had a meeting over there. I might as well tell you that our group is for you, for the head of this organization. Morgan & Hodges (Harjes) are against you. The Morgan interests say that you cannot be trusted, that you are too radical, and so forth, that you are too much on the side of the little fellow; you cannot be trusted. *They are for Douglas MacArthur as the head of it. Douglas MacArthur's term expires in November, and if he is not reappointed it is to be presumed that he will be disappointed and sore and they are for getting him to head it."*

I said, "I do not think that you will get the soldiers to follow him, Jerry. He is in bad odor, because he put on a uniform with medals to march down the street in Washington. I know the soldiers."

"Well, then, we will get Hanford MacNider. They want either MacArthur or MacNider. They do not want you. But our group tells them that you are the only fellow in America who can get the soldiers together. They say, "Yes, but he will get them together and go the wrong way." That is what they say if you take charge of them."

I said, "MacNider won't do either. He will not get the soldiery to follow him, because he has been opposed to the bonus."

"Yes, but we will have him in change (change?)"

And it is interesting to note that three weeks later after this conversation MacNider changed and turned around for the bonus. It is interesting to note that.

He said, "There is going to be a big quarrel over the reappointment of MacArthur" and he said, "you watch the President reappoint him. He is going to go right and if he does not reappoint him, he is going to go left."

I have been watching with a great deal of interest this quarrel over his reappointment to see how it comes out. He said, "You know as well as I do that MacArthur is Stotesbury's son-in-law in Philadelphia—Morgan's representative in Philadelphia. You just see how it goes and if I am not telling you the truth".

I noticed that MacNider turned around for the bonus, and that there is a row over the reappointment of MacArthur. So he left me saying, "I am going down to Miami. . . ."

There are other portions of the suppressed testimony such as Butler's story of the conversation he had in Indianapolis with a man named Flagg who knew all about the fascist plot to organize an army directed by Wall Street financiers. I have been unable to locate a man by that name who is an Indianapolis publisher, as he was introduced to Butler, and I am inclined to think that he was masquerading under a different name and had been sent there to feel out Butler. Because of my inability to locate any such person I am not quoting the testimony.

The most significant part of all this suppressed evidence is that the Dickstein Committee dropped it like a hot coal though there w[illegible]lenty of evidence of a fascist-militarist plot. Nevertheless, when the Congressional Committee had MacGuire on the stand repeatedly, it questioned him about his finances but not one single question was directed at him regarding the American Liberty League, controlled by the du Pont interests (which are tied up with Morgan interests and Morgan interests are tied up with Warburg interests and Warburg interests control the American Jewish Committee which in turn guided this Congressional body) nor of the discussions sworn to under oath about the Remington Arms Co. supplying arms and equipment for the fascist army.

Not a single question was directed at MacGuire regarding the conversation testified to by Paul Comly French. Not a single du Pont or Remington Arms official was called. No—not a single official of the Liberty League on whose body are members of the American Jewish Committee, such as former Judge Joseph M. Proskauer.

If the Congress of the United States really wants to investigate fascist activities why does it not ask this Committee why this testimony was suppressed?

Next week John L. Spivak will present details of how the Warburg-Morgan interests are tied up; and how the Warburg interests control the American Jewish Committee, leaders of which guided this Congressional Committee in suppressing the evidence of the fascist conspiracy by Wall Street financiers.—THE EDITORS.

THE INTRICATE STRUCTURE OF WALL STREET'S FASCIST CONSPIRACY

WALL STREET'S FASCIST CONSPIRACY

2. Morgan Pulls the Strings

JOHN L. SPIVAK

9

AS Wall Street moves toward a fascist dictatorship to head off the growing revolt of the people against continued hunger and misery in a land of plenty, its activities take in a far wider range than the purely military aspects of a seizure of power. The suppressed testimony of the Dickstein-McCormack committee, disclosed in THE NEW MASSES last week, dealt largely with the efforts made to induce Gen. Smedley D. Butler to assume the leadership of a military coup. Side by side with preparations for such an attempt goes the daily and hourly campaign of manipulating the mind of "the public" into a mood receptive to Fascism in the United States.

Three main agencies are at work in the fascization of public opinion: the press, the radio, and propaganda organizations set up for this specific purpose or converted to it.

... the press, of course, the most important ... is the sinister one of William Randolph Hearst; Father Coughlin and a host of would-be imitators of the priestly demagogue fill the air from morning to night with radio appeals to exterminate the Reds; and the American Liberty League, and the Crusaders, are outstanding examples of the organizational aspects of the fascist conspiracy.

All these activities go on simultaneously, and to some extent, on a superficial view, independently of each other. It will be shown in this article that there are organic links tying the various branches of Wall Street's fascist conspiracy together. An understanding of the financial set-up behind the propaganda for Fascism, as well as behind the more spectacular moves for a fascist army, is essential before the effect of such propaganda can be successfully combatted. The buyer of a capitalist newspaper who reads a violent assault on the Communists, and then listens to an equally bitter tirade against "the Reds" on the radio, should know that there is a definite connection between these two pieces of fascist propaganda. There are rifts and contradictions within the capitalist camp itself—Morgan against Rockefeller interests, to name the main one—but on this question of Fascism vs. Communism Wall Street works as a unit.

In the open attempt to get Gen. Smedley D. Butler to organize 500,000 war veterans ... fascist army we can trace the financial backing and from there continue on to the interlocking financial interests of those behind the fascist conspiracy and the reasons for the suppression of evidence about it by a Congressional Committee instructed to seek just such evidence.

Gerald C. MacGuire, the $100-a-week bond salesman who approached Butler with the suggestion, was ostensibly working for Robert Sterling Clark, who inherited the Singer Sewing Machine millions. During the period that MacGuire was seeking a leader for the proposed fascist army, he was kept on the payroll of Grayson M-P Murphy, a Wall Street broker. In his testimony before the Dickstein Committee which was suppressed, MacGuire said to Butler:

> The Morgan interests say that you cannot be trusted. . . . They want either (Douglas) MacArthur or (Hanford) MacNider. . . . You know as well as I do that MacArthur is Stotesbury's son-in-law in Philadelphia—*Morgan's* representative in Philadelphia. . . .

The man named by MacGuire as having written the gold standard speech, for delivering which Butler was offered an $18,000 bribe by MacGuire, is John W. Davis, chief Morgan attorney, and one of a select few on the Morgan "preferred lists"—friends of the firm who are offered stocks or bonds at the original price of issue regardless of the market price. Say a stock was originally issued at $20. It is quoted at $36. Those on the Morgan preferred list are offered 1,000 shares at $20. They can sell it the same day for the market price of $36,000—making a neat little profit of $16,000. That is the way money is disbursed by the Morgan crowd to its favored few.

Murphy and Morgan

Grayson M-P Murphy, like John W. Davis, is one of the favored few on the Morgan preferred list.

Murphy is known in Wall Street as a Morgan man.

Murphy is a director of the New York Trust Co., a Morgan bank. A bank is known as a Morgan institution when one of the Morgan partners is on the board of directors. On the New York Trust there are two Morgan partners: A. M. Anderson and H. P. Davison.

Murphy is a director of the Guarantee Trust Co., on which two more Morgan partners are directors: Thomas W. Lamont and George Whitney.

Murphy put up $125,000 to organize the American Legion, which has functioned openly many times as a strike-breaking organization.

So much for Murphy's history for the time being. We will return to him.

John W. Davis, once in the field for the Presidency of the United States, is Morgan's chief attorney. When a Senate investigating committee tried to get income tax reports of the world's leading private banking house, this man who wanted to be President of the United States bitterly fought every move designed to reveal its income.

Davis is one of those on the Morgan preferred lists.

Davis has borrowed money from the Morgans.

Davis is a director of the Guarantee Trust Co. of New York—the same bank that Murphy is a director of and which has two Morgan partners on the board of directors.

Davis is the man who was named in Butler's testimony as the one who wrote the gold standard speech which MacGuire tried to bribe Butler to make at the American Legion convention.

Davis' name was suppressed by the Dickstein-McCormack Committee.

Davis was never questioned, either in a hearing or by a Committee investigator, whether he ever wrote that speech.

The reader is asked to bear these italicized facts in mind, for the financial tie-ups of the Wall Street financiers and their fascist activities are complicated.

During the period that MacGuire, Murphy's employe, was maneuvering to get Butler to make the gold standard speech as well as organize a fascist army with a guaranteed backing of an initial $3,000,000 and a promise of $300,000,000 more if necessary, this bond salesman who had never had more than a few thousand dollars to his credit in any bank suddenly began to make amazing deposits. Let me quote a few from his bank record. At the Irving Trust Co., listed in his own and his wife's name: G. C. MacGuire and Elis. W. MacGuire "or either of them" of G. M. P. Murphy & Co., 52 Broadway, New York, he made these deposits

FEBRUARY 5, 1935

GRAYSON M-P MURPHY
Morgan's man, who kept Gerald C. MacGuire on his payroll

GERALD C. MacGUIRE
Who tried to get Gen. Smedley D. Butler to organize a fascist army

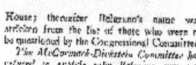

GEN. SMEDLEY D. BUTLER
Who turned down this particular offer of a ride on a white horse

On July 27, 1934, he deposited $39,306.78.

On July 27, 1934 (the same day) he again deposited $97,766.94.

On August 24, 1934, he made four deposits as follows: $31,222.92; $31,128.36; $05,-365.39, and $25,643.35.

Two days later, on August 16, he again deposited $45,805.56.

Before this sudden splurge in high finance MacGuire's normal deposits were less than $100 and occasionally $250 or $500, as commissions for selling bonds. Never did he deposit sums anywhere approaching these considerable figures quoted above.

In accounting for his expenditures during the American Legion Convention in Chicago in 1933 to the Dickstein Committee, this seeker for a fascist army tried repeatedly in trying to account for an expenditure of $53,-000 which he could not account for.

The Dickstein Committee never cited him for contempt or called his employer, Murphy, to explain what all these high financial transactions meant.

Belgrano and the Legion

Murphy has had his fingers in the American Legion pie ever since he first advanced the $125,000 to organize it. The Legion has been in the control of a small clique, but suddenly at the last convention the clique was overthrown and a California banker, Frank N. Belgrano, put in as Commander.

Belgrano was summoned to Washington to testify before the McCormack-Dickstein Committee about his knowledge of American Legion activities. Before he was called to the Committee room, Belgrano had a private conference with President Roosevelt at the White House; thereafter Belgrano's name was stricken from the list of those who were to be questioned by the Congressional Committee.

The McCormack-Dickstein Committee has refused to explain why Belgrano was not called.

In this connection Belgrano's financial interests become important.

Belgrano is associated with the Giannini financial interests on the West Coast, and more specifically with A. P. Giannini in the Trans-America Corp.

One of the directors of the Trans-America Corp. is Elisha Walker, a Kuhn-Loeb partner.

Giannini is tied up in more financial ventures on the West Coast with a gentleman named William Randolph Hearst, and is known in California as a Hearst man.

The Giannini banking interests are closely tied up with the National City Bank where both Rockefeller and Morgan interests merge. When Charles Mitchell was thrown out of the presidency of this world-powerful bank, it was known in Wall Street that President Roosevelt had recommended Perkins as the new head because the President knew him well through the Roosevelt family holdings in the Farmers Loan and Trust Co.

In this connection I should like to call the reader's attention to two more bits of testimony suppressed by the Dickstein-McCormack Committee:

Paul Comley French, reporter for the New York Post and Philadelphia Record, swore under oath:

> He (MacGuire) suggested that the General organize this outfit himself and ask a dollar a year dues from everybody. We discussed that and then he came around to the point of getting outside financial funds and he said that it would not be any trouble to raise a million dollars. He said that he could go to John W. Davis or Perkins of the National City Bank, and any number of persons and get it.

And Robert Sterling Clark told Butler, according to the suppressed testimony:

> He said, "You know, the President is weak. He will come right along with us. He was born in this class. He was raised in this class, and he will come back. He will run true to form. In the end he will come around. But we have got to be prepared to sustain him when he does."

Wall Street's Crusaders

Let us now consider some of the leading fascist organizations and see where the Wall Street financial trail leads.

There is an organization known as the Crusaders with national headquarters in Chicago which has been broadcasting on a national hook-up twice a week and carrying on an intensive propaganda campaign of printed matter. The Crusaders were originally organized to fight for the repeal of the Prohibition Amendment. At that time Wall Street financiers, fighting against increasing taxation due to unemployment relief legislation, thought that by giving the people beer and liquor they could get their own taxes reduced. The Prohibition Amendment was repealed and the Wall Street interests which had backed the Crusaders, instead of letting the organization die, decided that so well organized a body could be used. It was next used in matters involving monetary changes, which was Wall Street's particular problem at the time. When the inflationists won out and the 59-cent dollar was established the Crusaders were then used as a political and anti-labor body, this being the first time that this

FELIX M. WARBURG
Who virtually controls the American Jewish Committee

SAMUEL DICKSTEIN
Whose "investigation" of Fascism was steered by Warburg and his financial interests

JOHN W. McCORMACK
Chairman of the "investigating" Committee which suppressed testimony on Fascism

organization, first organized to fight prohibition, began openly to participate in politics and anti-labor moves.

Its first active work was to help defeat Upton Sinclair for the governorship of California. The California financiers and industrialists were afraid of the effect of a victorious campaign based, as Sinclair's was, on the deep-seated discontent of the masses of the people. The Crusaders jumped into the fight on the air and in disseminating countless thousands of leaflets and throwaways. Samples of the sort of propaganda they issued can be seen in their comment upon a meeting of unemployed. Their headline was:

UNEMPLOYED HORDE FACES CRUSADERS' OPPOSITION

In other propaganda they warned that "The California Crusader Rides Again!" Most of their warning was against "radicalism" and the rising militancy of the California workers.

Today the Crusaders following pretty much the same tactics, keep the sources of their income a deep secret; it thus becomes important to see who are their active supporters:

There is John W. Davis, Morgan's chief attorney.

There is James P. Warburg, of the Kuhn-Loeb Warburgs.

I name just these two to show the tie-up of Morgan and Kuhn-Loeb interests when it comes to supporting a fascist body actively participating in anti-labor moves. These two represent apparently opposing financial interests; as well as different racial and religious groups, both of which are now working together in promoting a fascist organization like the Crusaders—an invaluable help to capital in carrying propaganda against militant labor.

The Crusaders are still in the organizational stage and are trying to get 10,000,000 members. They are extremely active among military men and in military schools. Most members do not know that one of the organization's chief functions is "fighting subversive elements," particularly in the schools. In charge of this branch of their activities is Col. Roy Felton Farrand, head of St. John's Military Academy, Delafield, Wisconsin, a member of the Crusaders' Advisory Council.

Other Crusaders

Other Crusaders on school and college boards are also using their influence "to protect the country from subversive elements" like Fred I. Kent of New York, vice-president of the Morgan-controlled Bankers Trust Co. Kent is President of the Council of New York University where student anti-military movements have been rigidly suppressed.

Kent also is an officer of the U. S. Chamber of Commerce which recently issued a broadside against "subversive" elements and particularly against elements which are trying to organize workers.

Wallace McK. Alexander, of San Francisco, of the Crusaders' National Advisors. Alexander is a sugar planter, a big business man and a trustee of Stanford University, where an anti-radical drive was started by the faculty after the formation of the Crusaders.

Sewell L. Avery of Chicago, a director of the Morgan-controlled U.S. Steel, is one of the Crusaders' National Advisors. Avery is a trustee of the University of Chicago where an anti-radical drive took a sudden spurt after the Crusaders went into active anti-labor activity.

Francis B. Davis, Jr., of New York, right hand man of the du Ponts, a director of the firm of the Morgan-controlled New York Trust Co.

Cleveland E. Dodge, of New York, vice-president of the Phelps Dodge Co., big copper producers, a director of the City Bank Farmers Trust Co. (where he ties up with the Hearst-Anaconda copper interests), Crusaders' National Advisory man, is chairman of the Board of Teachers College, Columbia University. The Dean of this college, Thomas Alexander, has been working with an agent of Royal Scott Gulden, head of the anti-semitic secret espionage Order of '76.

Albert D. Lasker, a member of the Crusaders' Advisory Council is a member of the American Jewish Committee.

Alfred P. Sloan, Jr., President of General Motors Corp., a director of E. I. du Pont de Nemours & Co. and other Morgan-controlled industries.

These few directors of this fascist organization are, I think, sufficient to show who are back of this organization and the effects of its activities.

Let us now consider the American Liberty League, an organization fighting for "respect for the rights of persons and property as fundamental to every successful form of government" as well as the "retention"—and the re-establishment where necessary—"of the American traditions of government and individual liberty."

This organization is especially strong on upholding Constitutional rights. I have never yet come across a fascist organization which did not base its righteousness upon "upholding the Constitution." The Liberty League, with almost one million dollars in its treasury as

WILLIAM RANDOLPH HEARST
Red-hunter-in-chief, boss of the American Legion, who favors the fascist plot

J. P. MORGAN
Ultimate fountain-head of the whole fascist conspiracy of Wall Street

JOHN W. DAVIS
Chief Morgan lawyer, shown to be tied up with fascist organizations

many more millions available if necessary, is now on a campaign to get 4,000,000 members—a powerful band if properly financed by the country's leading financiers.

Among the heavy backers of the American Liberty League (the name of which was suppressed by the Dickstein Committee when Butler and French testified about it) are:

John W. Davis, Morgan's chief attorney.

Sewell L. Avery, of Chicago, a Crusader advisor.

W. S. Carpenter, Jr., of Wilmington, one of the original organizers of the League, tied up with the du Pont-Morgan interests.

Robert Sterling Clark of New York, who gave large sums of money to Gerald C. MacGuire, the $100-a-week bond salesman, to find a leader for a fascist army.

Irenee, Lammot and Archibald du Pont, all of the duPont munitions family.

There are others, but these are sufficient to see who is organizing a body to "re-establish where necessary" the "respect for property."

Liberty League officials who are on the National Advisory Council of the Crusaders are:

John W. Davis, Morgan man.

F. B. Davis, Jr., du Pont right-hand man.

Sewell L. Avery, the big business and education man from Chicago.

One of the original founders of the American Liberty League is Jouett Shouse, at present president of the League. Shouse married the daughter of Filene, partner of Kirstein, who is on the executive Committee of the American Jewish Committee.

Joseph M. Proskauer, former Appellate Court Judge of New York, a director of the American Liberty League, is a member of the Executive Committee of the American Jewish Committee.

Leaders of the American Jewish Committee steered the work of the Dickstein Committee in its investigation and helped direct the authorities' publicity. The Dickstein Committee suppressed Gen. Butler's testimony regarding the American Liberty League.

John J. Raskob, du Pont and Morgan man,

And— Grayson M-P Murphy, the Wall Street broker, who kept MacGuire on the payroll while the latter was trying to get Butler to organize a fascist army.

There are other organizations similarly tied up with powerful financial cliques like the Committee for the Nation and Father Coughlin's Union for Social Justice, the details of which would only clutter up a story of the interlocking financial powers behind the moves to establish Fascism in this country. I shall mention only one or two individuals and their relations to the financial powers behind the Dickstein Committee and its suppression of evidence of Fascism.

One of the leading members of the Committee for the Nation is Lessing Rosenwald, of Sears, Roebuck. Rosenwald is on the Executive Committee of the American Jewish Committee. Rosenwald worked closely with Edward A. Rumely, secretary of the Committee for the Nation, whose name never appeared on the Committee's letterheads. Readers of THE NEW MASSES series on the Nazi activities in the United States will recollect that Rumely is the mysterious gentleman who gave Viola Ilma letters of introduction to Nazi leaders in Germany and that the Committee by putting across inflation in the United States saved Germany millions of dollars since Germany had the largest floating debt here of any country.

Father Coughlin is tied up with the Rockefeller interests as well as with Ford and Hearst and is not so guileless in his attempts to organize a fascist army under the guise of a Union for Social Security. The monetary reforms which he is advocating are of great help to the Rockefeller interests, Ford and Hearst.

Let us now consider William Randolph Hearst and his current efforts to scare up the "Red" bogey as one of the first steps in preparing the country for Fascism. Hearst with his chain of newspapers reaches millions of readers. Just before he started his anti-Red drive he returned from a visit to Germany where he had conferred with Hitler and other Nazi leaders. Shortly after his arrival home he stated in a front page editorial that this country need not fear Fascism, that Fascism can come only when a country is menaced by Communism.

Mr. Hearst has about a quarter of a million acres of land on which he grows vegetables and fruits. California migratory workers work for him as they do for other land barons. For years the approximately 100,000 migratory workers in that state had wandered from farm to farm working from sunrise to sunset,

...and earning scarcely enough to buy food for themselves and their families. The American Federation of Labor had never tried to organize these workers because they did not pay dues. The Cannery and Agricultural Workers' Union, affiliated with the militant Trade Union Unity League, did organize the migratory workers and did win an overwhelming majority of the strikes they led. The cry "Communist agitators" sounded all through the state. Then came the great general strike last summer. Again the cry "communist agitators" was raised. It was Hearst's attorney, Neilan, working with A.F. of L. men, who helped break the strike by creating a split in the ranks of the strikers.

Hearst owns the Homestake mine, one of the world's largest gold mines. When inflation came and your purchasing power was reduced by a 59 cent dollar the price of gold went up. Hearst made millions of dollars. The Committee for the Nation, steered by Rumely, close friend of Nazi leaders, brought about the inflation.

Hearst has tremendous copper interests, especially the Cerro de Pasco Copper Corp. Hearst's copper interests tie up with Morgan and Kuhn-Loeb financial interests. At the time Hearst opened his barrage against the Communists in the United States, a Nazi commission was about to leave Germany for this country to try to arrange credit of millions of dollars worth of copper and other war materials. Hearst's copper was among those the commission had under consideration. Among those on the Nazi commission were directors of the Vereinigte Stahlwerke, the German steel trust which originally gave Hitler money to organize his fascist army, and Berlin bankers. Visiting Germans included:

Albrecht von Frankenberg und Ludwigsdorf,
Dr. Oskar Sempell,
Hugo Stinnes,
Dr. Georg Solmssen, banker of Berlin,
Prince Gottfried zu Hohenlohe von Langenberg of Berlin,
Ernst Wallach, Berlin banker.

Present in the United States at the same time was Max Warburg, brother of Felix Warburg who controls the destinies of the Kuhn-Loeb financial empire. Max was interested in the "stand-still" agreement with Nazi Germany which would result in the bankers getting their interest on short term loans, of which the Warburg-controlled Bank of Manhattan held millions of dollars. This would enable Hitler's Germany to get credit.

Hearst's Tie-ups

Hearst, too, is tied up with the *Morgan-du Pont* interests in the Marine Midland Trust Co.

Hearst is tied up with the Giannini interests, and *Warburg partners are similarly tied up with Giannini interests.*

Giannini controls Frank N. Belgrano, present Commander of the American Legion, which the Morgan interests want to use as a base for the fascist army.

Belgrano is at present cooperating with Hearst in his red scare propaganda.

Hearst's financial man, Edward H. Clark, is on Seaboard Oil and Seaboard Oil has joint interests in oil properties with the Royal Dutch Shell. The Royal Dutch Shell is controlled by Deterding, who backed Mosley's English fascist army financially.

Hearst is tied up in Canada paper with Rothermere, who also backed Mosley's fascist army in England.

Samuel Dickstein, vice chairman of the Congressional Committee which suppressed testimony of Wall Street's fascist plot, is known as a Hearst man in Washington with Hearst papers playing up almost everything that the Congressman utters, particularly against the Reds.

Warburg's Tie-ups

Let us now consider the Warburg financial tie ups, the reader bearing in mind that leaders of the American Jewish Committee, virtually a Warburg-controlled body, directed the activities of the McCormack-Dickstein Committee.

Felix Warburg is a director of the Morgan-controlled American Securities. Another director on this Morgan-controlled corporation is Walter Frew. Frew is one of the men who gave money to Gerald C. MacGuire while the latter was trying to organize a fascist army.

Warburg is a director of the Bank of Manhattan and virtually controls it. In this bank worked F. X. Mittermeier, a secret Nazi agent.

The Bank of Manhattan, Warburg-controlled, is one of the largest holders of German short-term notes in the country.

Kuhn-Loeb underwrote the North German Lloyd Line and the City of Hamburg, where Felix's brother Max is a Jewish banker getting along very nicely in a land where poor and middle class Jews are being killed, tortured and driven into ghettoes.

Lewis L. Strauss, a partner in Kuhn-Loeb, is a director of the Morgan-controlled New York and Susquehanna R. R. and the United States Rubber Co., the latter of which is controlled by the du Pont-Morgan interests who are the leading figures in organizing the American Liberty League and giving support to the *Crusaders.*

Sir William Wiseman, a Kuhn-Loeb partner (formerly head of the British secret service in the United States during the World War) is a director of U. S. Rubber, du Pont-Morgan controlled and of the Morgan-controlled National Railroad of Mexico.

Elisha Walker, a Kuhn-Loeb partner, is a director of the Morgan-controlled Radio-Keith-Orpheum, the Giannini-controlled Bank of America, Transamerica Corp., General Foods, and Bancamerica-Blair.

When I say Giannini-controlled, I mean also Hearst-controlled.

I think these few Warburg partners' financial tie-ups with the forces at work in promoting Fascism in the United States will give the reader a clearer notion of why the Warburg interests are not fighting Fascism though if Fascism comes it will inevitably bring anti-semitism in its wake.

The Committee's Tie-ups

Let us now consider the financial tie-ups of the American Jewish Committee, leaders of which steered the McCormack-Dickstein Congressional Committee which investigated the Nazis a little, issued a line of propaganda against Communists and suppressed evidence of *Fascism.*

I point these out not as casting reflection upon the great body of sincere Jews cooperating with the Committee's work in fighting anti-semitism but to point out that the banker, whether he be Jew or gentile, is interested first in his class interests and in fighting for these interests he forgets racial and religious affiliations.

The American Jewish Committee has always opposed the boycott of German goods.

Kuhn-Loeb underwrote the North German Lloyd and the City of Hamburg and have millions of dollars invested in Germany.

Henry Ittleson is a member of the Executive Committee of the American Jewish Committee. Ittleson is president and director of the Commercial Investment Trust Aktien-Gesellschaft of Berlin.

Irving Lehman is a member of the Executive Committee of the American Jewish Committee. The Lehman brothers have large interests in Nazi Germany.

Albert D. Lasker is a member of the Executive Committee of the American Jewish Committee. Lasker is on the National Advisory Council of the Crusaders which got money from the American Liberty League.

Lessing J. Rosenwald is a member of the Executive Committee of the American Jewish Committee. Rosenwald is on the Committee for the Nation; Sears Roebuck gave money to the American Vigilant Intelligence Federation of Chicago, money which was used to disseminate anti-semitic propaganda.

Roger W. Straus is on the Executive Committee of the American Jewish Committee. Straus is a director of Revere Copper and Brass and is tied up with other copper interests in which the German commission to this country was profoundly interested.

Louis Edward Kirstein is a member of the Executive Committee of the American Jewish Committee, is vice-president of William Filene's Sons & Co. of Boston. Filene's daughter married Jouett Shouse, president of the American Liberty League.

Kirstein is a director of the Morgan-controlled Radio-Keith-Orpheum on which Warburg partners are directors.

Joseph M. Proskauer is a member of the Executive Committee of the American Jewish Committee. Proskauer is close to Hearst interests and is a director of the American Liberty League.

I have been unafraid for all

these years, as an activist; as a captain enjoying the inescapable eye of a great hurricane.

But with the daily flood of information, of the planet meltdown happening this year, has finally made me afraid.

I lay awake at night

thinking what about my grandchildren. Now I worry about my son.

It looks like we are in big trouble
and it is now not tomorrow
somewhere else in time.

I am finally afraid for the Earth

And our president and the military
have gone mad with another war
for MORE LIQUID SUICIDE

THE FINAL OIL BURNED

WILL SEEN BY FEW.

Oil, if burned, that the math says will kill all of us in the next 30 years. If you are a 'Darwin' then you may be the last to die or you will survive this extinction event and maybe learn a lesson. We are all teachers of evolution or not. Even extinctions have exceptions.

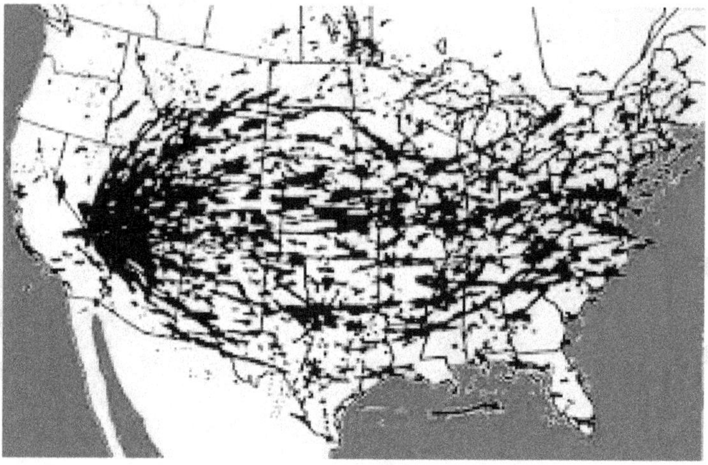

THE SIGNITURE OF MAD MEN

TRACKS OF ALL THE BOMBS USE IN THE 'EXPERIMENT'

IS A TOXIC CHEMICAL DEED THAT ALTERED THE MIND AND BODIES OF AMERICA FOR THOUSANDS OF DECADES TO GET THE MEDICAL BUSINESS RESULT THEY WANTED?

ARE THEY REALLY, CULLING US AND BETTING ON THE RESULTS, LIKE RACING PIGS IN A MACHIAVELLLIAN GAME?

YES, IT IS LIKE GERMAN

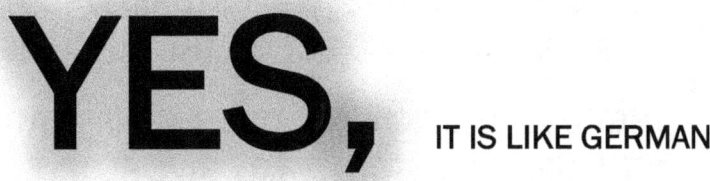

ROULETTE WITH EARTH AND OUR HEAD IS WHERE THE BARREL(S) ARE POINTED.

WHEN THEY: THE ROYALS, THE SELF CHOSEN, THOSE WHO ARE LAZY AND CONSIDER BUSINESS AS WORK, IT IS NOT WORK ARE CORNERED THEN THEY WILL LIKE IN 'ATLAS SHRUGGED" DISAPPEAR AND WE WILL BE BETTER.

LIVING UNDERGROUND FOR HUNDREDS OF YEASR WILL BE LITTLE FUN QUICK.

I MUST DEFINE THE PLANNERS: CORPORATISM OR FASCISM IS BUSINESS USED AS A "NECESSARY" MANIPULATION OF PEOPLE AND RESOURCES BY THE DICTAOUS.

AFTER CENTURIES OF PRIVALEGE AND POSSESSIONS THEY, THE NOW INBRED AND WEAKENED MORALLY, ARE BORED WITH WEALTH POWER, SEX IS GONE WEIRD, BEST DRUGS NO LONGER WORK WELL WITH HYBRIDS, SO THEY START WARS AND BET ON THE OUTCOME.

THEY USE CHEMICAL/WEATHER MANIPULATIONS WITH DIRECTED STORMS TO DRIVE PEOPLE AWAY FROM THE OIL. KATRINA WAS A DIRECTED STORM.

JULY 2019 ANOTHER DIRECTED STORM WILL EMPTY THE COAST OF MORE OLD DNA AND OTHERS WILL MOVE IN AND TAKE CONTROL.

POSSESSION IS THE RULE IN ALL TENSE. RECENT GOVERNMENT DEMOGRAPHICS STUDY SHOW THE INFLUX INTO THE OIL REGION, AFTER SEVERAL STORMS ALL DIRECTED WAS 35% GERMANIC. BRILLIANT IDEA!

HEY, I JUST READ THE STUDY AND A JEWISH FRIEND IN NEW ORLEANS TOLD ME RECENTLY THAT THE CITY HAD TAKEN ON A GERMANIC FEEL IN THE HIGH RENT DISTRICTS. FIND IT YOUR SELVES, LOOK IT UP

I REPEAT

IT SEEMS AS THE COAST WAS DISTRACTED AS REPORTED BY GOVERNMENT PUBLISHED MIGRATION REPORT, AND A NEW ORLEANS JEWISH AFFILATED ACTIVIST OBSERVATION THAT THE NATURE OF NEW ORLEANS TOOK A "GEMANIC FEEL"

Just do something; it could be a gift. Think your time is up and dislike that thought then act. Now the action must have a plan.

Just today July 4, 2019 I get a news report that KLM Dutch airlines was asking flyers to not fly and rethink their trip. The next day the French did the same thing

This begins the changes of the system and maybe the mind of the world away from killing the life support system. This is a ray breaking.

I WONDER IS THE LIGHT FINALLY COMING

These clouds on our horizon are dark but we are capable, with chemical caveats, of thinking our way out of this. Remember to remember this: That information was given on camera in 1987 in Rhode Island. I was President of and executive producer/producer for *Survival of the Sea'.*

We rented a home on Narragansett Bay there for a production office for a few months as we worked on 'Never A Sabbath'.

Peter Benchley a frequent working guest and he was Narrator. I was producer and co-writer and most else it took to make a major film that "looked where no one even thought was a place to look".

Metaphysics and dialectic logic. I did not give myself a choice.

Thank you, Walter Cronkite: I only recently feel the significance of a fisherman meeting and turning the camera on the media icon of America. And I did it very, creatively, well as a self-trained fisherman/filmmaker with the help of a few punishing mentors.

That icon for some reason felt silently that I was the man to focus this thought upon.

My camera man and Survival of the Sea board members Sister Francis Fisher, Fred Rothschild and Jaws author Peter Benchley heard it that day. But Walter was looking in my eyes from three feet away. He knew I was looking for answers for real. I had met him through Peter Benchley a few weeks before on Walter's sail boat in Newport.

I must have gotten the cosmic message. A seed thought was planted in my mind requiring, sub-consciously, an inquisition over time to discover what is at the end of this quest for understanding.

This process of seed thought growth grew into this and other books of mine and many movies seen around the world on television and Internet.

The understanding of seed thought growth came over time. Once I understood that, with me, information, or incomplete information and observations comes in and transmuted into complete thoughts. It, the brain, works information, like a computer doing calculation, into understanding. The one twist—you must power your imagination with ever cell and the mind will obey.

Through the power of imagination thoughts correct, are received and have been, religiously, passed on. I know some hear me and some thank me, hopefully. This reality is a lot to carry I am aging, still strong but I tire of the feeling I get that most people feel it is not something they need to know; wrong. But it works for me.

He helped me to a path of enlightenment that now has me understanding me and a mind changing planet reality.

I have advocated the idea of the challenge of an across the board halving of the fossil use in a matter of weeks. And not shift the fuels to those who now waste the most with wars and private jetting to eat caviar in another country. Then be back home in hours for drinks without regard for our planet.

Equality is upon us. The planet situation is very Equanimeous

During the decades, long process, of writing these life observations and assembling the works of fine minds, into a "connecting of the dots" I became aware of the pink elephant in the room no one mentioned.

Most everyone was by, kingdom decrees' 'redacted' first regarding education and history. Past, present and seeing future is easy. Then I saw Armageddon engineered as a business deal and rural restructuring of people. They see people as commodious units to be traded and balanced in the books.

Yes, the game is that big and these people are on life support for they do have a heart or conscience.

They had their chance and fucked-up

I see the white caps breaking over a final reef. Let me at the wheel metaphorically. I am not running for office ever but if I did my campaign slogan would be

Vote for me

The other guy

is worse

They finally breed a cabal of cross breds who do not care. If the devil is careless then the metaphor is live and I have been disproven about the images of the bible I was branded with in fire and Baptist brimstone.

Given the magnitude of these messages: A comic report was on the Seth Myer show, July 4 2019 making fun of Trump getting German bank money that he lost or secretly invested.

Is this the same game making presidents and building images by investing in shield organizations working for world domination connectivity purposes.

A Nazi corporate spawned banking system brought to America was the first of major action of infiltrators in 1917

The take-over of American resources by capturing the countries' economies with the federal reserve system was essential.

Now the idea that the **Jews** are the beast here **should be put aside.**

German infiltrators, hiding in the skirts of Judaism and a created by them anti-Semitism social trap for culture is the structure we should fear

It is too late to cry for long. this is on us everywhere as, biblical of proportions, Armageddon like storms and flooding everywhere and everywhere are planes flying high without interference seeding these storms in the battle of the bulging dikes and river banks; the attack with world hunger implications.

Germany won the war of the minds and came here mostly hidden as German Jews and captures America economy to use as they are using us now to terrorize the world.

It is a great day to kill some children was a poem of the Persian war number 99999999. Been going on forever it seems.

Trump is being funded by a bank formed out of corporatism building efforts to collect from both sides of the same ball of war.

it is their game. It is so big and so sinister that to be able to visualize, without corrupting the mind this creepy shit, is the challenge. This is straight out of the loony stars.

Yes, this electoral coup Trump is being funded by the air money planned business acquisition of the world. Control without regards business deal that has ruined earth.

May they rest in peace in those undergrounds cites. Their generational misery of living underground with be 10x10x10x10 on and on of reflection of how wrong and evil this true story is and know it is the mystery of the pendulum that will deliver them fresh misery for these deeds.

The universal pendulum at the human or life planet level when balanced should be at the top of the fulcrum of the pendulum and the swing of things is seen and not ridden.

WE have let, and participated in this cycle of greed story way too long and we are now about to evolve and the 'Darwins' will survive.

It may be those going underground to survive the fall of the ecosystems of Earth and maybe the seesaw will come out Even underground the fittest will be the survivors and they may have learned to appreciate earth: the life-giving organism we were given to vacation upon and go back and add good intent to the world of thoughts stored in the magnetosphere or our hard drive filled with planet memories of evolution. As an astral body being, **AGAIN** while enjoying mega-nature and its lights absorb and become a star and shine on the all of future earth. Yes, we are

ONE OF MANY GO(O)DS. THIS IS A UNIVERSAL REALITY OF PHYSICS. HOW YOU USE YOUR PHYSICS IS YOUR DECISION(S)

SO, SO SAD ABOUT THIS MESS
MESSENGER THING

SORRY I MUST BE THE ONE

TO SHOW YOU THESE PROFOUND
TRUTHS

WHY ME? WELL, WHY NOT?

I DID IT MY WAY, EVEN THE SCEW-
UPS AND I LIVED ALMOST

COMPLETE

I EXPERIENCED MANY GOOD FEELINGS AND LAUGHED A LOT ALONG THE WAY

AND I HAVE BEEN CALLED A HERO FOR DOING THING THAT NEVER FELT HEROIC

Elections

Are

A rig-able

Formality

In a fooled

Democracy.

The Millenium Project

The Millenium Project is a White House-sponsored call to communities across the nation to celebrate the year 2000 with new ideas for improving the social condition.

Capt. Gary Burris answered the call this month with the following letter sent to President and Mrs. Clinton. He proposes a Green Model for Key West that demonstrates the ways our island could become increasingly self-sufficient, and thereby serve as a model of radical interdependence for the rest of the country. His letter was accompanied by endorsements from Jimmy Buffett, who is known and admired by the Clintons, and the principal of the high school that both Bill Clinton and Burris attended in Arkansas.

Cognizant of how accessible Clinton's White House presents itself — call (202) 456-1414 and a live human being will talk with and listen to you, evenings and weekends included — *Solares Hill* is curious as to how well Burris' proposal will be received and supported. Frankly we'd be surprised if this clear-eyed approach to the possibilities of a better life in Key West goes unrecognized. We will update you.

M.H.

> I would rather die in the quest
> than explain the failure.
>
> I would rather be courageous
> than stable.
>
> I would rather be known for
> finishing late
> than for not finishing at all.
>
> — From "The Olympic Runner"
> by Gary Burris

Dear Mr. President and Mrs. Clinton:

I recently read of your Millennium Project, and I knew it was time to write and inform you of something special developing in the Florida Keys.

We call it the Green Model. In a nutshell the goal of the Green Model is to create a working community model of environmental responsibility (both urban and rural). It could be called a system-wide experiment using proven technologies to build a human system that better functions in unison with the natural system. There are components of this model that reduce the release of some gases harmful to the atmosphere.

This Green Model idea had its genesis in our home state of Arkansas. Ironically, among the men who sat and listened to me speak of ideas to defend the oceans and the environment were men who had a lot to do with your early political encouragement — Alphonso Logan and Arthur Wimmell. We must have gotten a lot of the same advice. Their advice to me was "go see Bill." You and I never connected back home, for I was either at sea, fishing commercially, or on the environmental discovery path, or you were busy building a better state.

Today, as I think back to the beginning of my commitment to a better world, I do know that even in the beginning I spent little time wringing my hands with worry for the planet. I approached negative environmental reality (after I got over the shock of realizing how bad it could get) with pragmatic thoughts for recovery. As I see it, the damage we have inflicted on the earth has a rip pattern, kind of like a tear in a shrimping net. To fix the environment, like a net, we must sew in reverse. We must turn around and retrace the steps to decimation. That is the dream of a Green Model — to build a pragmatic and profitable return-road to recovery.

I ask you to consider including the ideas and example of our Green Model into the plans for the Millennium Project. Especially include our home port of Key West as a Millennium Festival Site.

This is the perfect place to construct a Millennium exhibit to show the technology in action that will change or lessen our negative impact on our life support system.

Our Green Model has several components. The major one is the use of alternative biofuels that lessen our impact on the environment. The great advantage of using these fuels is the reduction of impact on the atmosphere. Another advantage is that the carbon dioxide released in the burning of vegetable-based fuels fits the ability of the planet to use this more "natural carbon."

An extraordinary benefit derived from development of this fuel to its highest potential is the fact that it gives us a way out of our addiction to harmful planet-damaging fossil fuels — sort of a light at the end of a smoky tunnel. There is enough waste vegetable oil available for recycling in North America to power the basic services needed to keep our system moving in the event that crude oil becomes scarce. And if we planted

Please turn to page 12

lem, so maybe you can help us convince our leaders that flooding the bay with dirty agricultural runoff only kills our downstream coral reefs.

But we must also look in our own backyards

must if we value our lifestyle, clean oceans and a healthy coral reef for the Florida Keys.

DeeVon Quirolo
Project Director, Reef Relief

MILLENIUM PROJECT

Continued from page 10.

the ground between the Interstate road systems of our country with oilseed plants, we could harvest many millions of gallons of clean usable fuel oil. Underlying all this is the knowledge, sad but true, that a sustainable system must be a profitable one.

Our ship, an historic North Sea fishing vessel named the *Dirk Seaguard*, will be in Chesapeake region during the coming months and we will be stopping in Washington D.C. to demonstrate an example of practical environmentalism. We extend an invitation to you to join us and see and smell the difference.

I have told many people that your administration would do something significant for the environment if not special-interested to death. The ideas we have to share seem to me, after studying them for many years, to be the most logical attempt available. I have also said many times that I should have at least one opportunity to share these solutions with my hometown President. I am glad you have this chance to tively effect our way of dealing with the Ea

Captain Gary Burris
Seaguard / Survival of the Sea Soci

UNITED STATES DEPARTMENT OF COMMERCE
National Oceanic and Atmospheric Administration
Rockville, Md. 20852
NATIONAL MARINE POLLUTION PROGRAM OFFICE

July 10, 1987

Mr. Gary Burris
Chairman of the Board
Survival of the Seas
3299 K Street, N.W.
6th Floor
Washington, DC 20007

Dear Mr. Burris:

Thank you so much for your help/ideas in formulating the municipal waste water working group. After a shakey start I feel we gunned down the most important aspects and fully expect to see your ideas reflected in the upcoming 5-year Federal Plan.

Good luck in your efforts on behalf of the marine environment. I hope to see you next time.

Regards,

Irwin Haydock

Irwin Haydock

Enclosure

Former fisherman sets sail to promote environmental causes

By MARK MUELLER
Journal-Bulletin Staff Writer

NEWPORT

NEWPORT — The ship's deck still bears some rust, and the 90 tons of riveted black iron does move ponderously through the water, but the 61-year-old vessel is afloat, as is the dream that has motivated Gary Burris for the past decade.

Burris, a former commercial fisherman turned environmentalist, has spent the past few months sailing the flagship for his organization Seaguard to East Coast ports, spreading his environmental message.

"If you educate people — give them the tools — they will take the initiative," Burris said.

With this doctrine, he and his vessel, the 85-foot Dirk Seaguard, have visited 16 ports in eight states and the District of Columbia. At each stop, the ship's crew performs several skits and a puppet show to educate people, especially children, about threats to the environment.

Burris founded Seaguard, based in West Kingston, two years ago and incorporated it six months ago. It is a grassroots organization that promotes environmental education and the need for regional oil spill response teams.

During the past few months, since the ship's restoration, Burris said more than 200 people have volunteered to help the organization — whether it's scraping rust off the ship's deck, or handing out literature and selling Seaguard T-shirts at various ports.

When not on tour, Burris keeps the boat in Newport Harbor, where he receives a free mooring.

Burris, 40, is an Arkansas native who came to Rhode Island about four years ago representing the Survival of the Sea Society, which he founded in 1980. Burris said the society deals more with producing educational films and lobbying for money and legislation, while Seaguard works to cooperate with established organizations. "It could be Save the Bay Seaguard," Burris said.

One of Seaguard's main concerns, Burris said, is preserving the oceans, which he says are the world's last great resource.

"We know that there are growing problems with plankton," which provides oxygen, Burris said. At the rate that trees are being destroyed across the world, the oceans will soon be the last supplier of oxygen, and to destroy the ocean is akin to suicide, he said.

He said that since the Exxon Valdez spilled 11 million gallons of oil off Alaska in 1989, an additional 10,700 spills of varying sizes, have been recorded. Oil spills also kill plankton, he said.

He said that other concerns, such as the greenhouse effect, are little understood by the public.

"The information is out there and has been out there for decades," Burris said. "The government just doesn't put the money into awareness.

"We want to teach the basics of how an ecosystem works. We couldn't wait for the government to do it," he said.

The campaign to create oil spill response teams to function "exactly like a volunteer fire department," has met with success so far, Burris said, with groups from [illegible] regions saying they will participate. In Rhode Island, eight people have signed up, and as soon as about 30 people — the minimum for a class teaching oil spill response techniques — join, Burris says he will go to Congress and sponsors to try to get grants.

Another Seaguard program, a film about the ecosystem and an accompanying magazine, will debut in Rhode Island schools this [illegible]. The program was scheduled to begin two years ago, but has been delayed because the federal Environmental Protection Agency said Burris's view of the environmental p[illegible]

SETTING SAIL: The 61-year-old, 90-ton, Dirk Seaguard plies its way through Newport Harbor.

Turn to SHIP, Page [illegible]

—Journal-Bulletin Photo by FRIEDA SQUIRES

Rainy day reading

a.. who is working at The Landing restaurant in Newport during sum-
Vermont, reads a book. Wet weather slowed yesterday's business.

Ship

Continued from Page F-1

ture as depicted in the film was unduly bleak.

But Burris says the film has been toned down and will be seen in Grades 8 to 12. "The film is a correlation between how the human body works and how the ecosystem on the planet works," he said, adding that it is tentatively titled "I Am Earth, You Are Earth."

Burris said that four new films are now in development, and the Dirk Seaguard soon will be headed for Belize, where Burris is instituting a program based on the one in Rhode Island schools.

Meanwhile, Burris still has work on Dirk Seaguard, as the restoration is three-quarters complete. Built in Holland in 1929, the ship, then named the Dirk Von Texel, was used to fish for herring and shrimp.

It was refitted as a recreational vessel in 1982, its holds converted into a galley and its sleeping quarters modernized. But during a heavy storm in Newport Harbor in 1987, the boat sank after it broke free of its mooring and was driven onto rocks.

Burris bought the boat for $20,000 two years ago, and has been working on it since.

"It was a wreck when we got it," he said. During the restoration effort, he has invested more than $100,000 in the boat and related projects, he said.

Once it was seaworthy, he donated the ship to Seaguard. But it still needs showers and plumbing, a new cabin, and the deck must be painted, Burris said.

The hull is now blue, and 30 small black flags adorn a line reaching to the top of the mast. The American flag flies at half mast. He said each black flag signifies the 1,500 acres of wetlands that are lost each day to erosion, development, highways and farming.

"We have the American flag at half mast all the time," Burris said. "Our country is dying, and the flag is at half mast."

May 27, 1988

WWOR-TV

Ms. Anne Wallace
Jesse B. Cox Charitable Trust
100 Franklyn Street
Boston, Mass. 02110

Dear Ms. Wallace:

My acquaintance with Gary Burris dates from November 1984 when he first approached Cable News Network in Atlanta with a fully conceived program about the decline of fisheries along the coastal United States. Ted Turner, WTBS and CNN Chairman and I, then the Executive Producer of CNN "SPECIAL ASSIGNMENTS" and Vice President for CNN, agreed the idea was well worth pursuing. CNN spent the next year filming and producing "Coastline Crisis" and in most respects this program did conform as Gary had visualized it. You may be interested to know that the program went on to win six national and two international awards.

As Executive Producer on that project, I had ample opportunity to observe and deal directly with Gary throughout the year that we worked together. His work on the project as a field producer on loan to the Network from the Survival of the Sea Society was invaluable. Without qualification I would characterize Gary as an exceptionally talented field producer, researcher and now a Producer-Director who, given adequate tools and resources, is capable of creating a program that will make a unique contribution and will conform to the highest broadcast standards.

Gary's intimate understanding of his subject matter, his real concern for environmental issues and his desire to make a difference keep him focused and striving for the best possible product. He also has a keen sense of balance in presentation of issues and the ability to explain arcane and complex information in ways that are relevant and entertaining to the audience. I have had occasion to keep in contact with Gary over the past several years and find that he continues to be current on environmental matters--in fact, more often than not, he is ahead of the times and is an invaluable source of information. He is currently advising WWOR-TV's Investigative Unit on a project related to the environment.

Gary Burris-2

I hope your foundation will take advantage of this opportunity to work with Gary and his organization. The project is timely, and given Gary's love of the sea and ecology and his skill with television, I am sure it will be a success.

Sincerely yours,

Ted Kavanau
Vice President News & Information
MCA/WWOR-TV

United States Department of the Interior

OFFICE OF THE SECRETARY
WASHINGTON, D.C. 20240

TAKE PRIDE IN AMERICA

Captain Gary Burris
Survival of the Sea Society
1863 Matunck
School House Road
Wakefield, Rhode Island 02879

Dear Captain Burris,

I enjoyed meeting you during my recent trip to Rhode Island and participation in the educational video taping for the "A BETTER BAY THROUGH BETTER EDUCATION" project. I understand that you may be using some of the video footage to produce public service announcements and a "Survival of the Sea" television program. We would appreciate knowing the outcome of these exciting possibilities.

I firmly believe that stewardship of our public lands and waterways is every American's responsibility. Your tireless efforts to promote proper care of the environment are both needed and greatly appreciated.

Thanks for all you are doing.

Sincerely,

Don Hodel

DONALD PAUL HODEL

BIOFUEL

From Page 1

more like popcorn than french fries in his skiff, which burns 100 percent Biofuel using a Yanmar diesel outboard.

"The best part is, engines last longer with it," Burris said. "It has no carbon in it. Pure carbon is an abrasive on the microscopic level."

Burris said Sebago is going to test the engine-reliability claims by running straight diesel in one engine and a Biofuel blend in the other in their two-engine catamarans.

Burris' biggest problem is persuading fuel dealers to stock Biofuel. "They tell me they're contractually obligated by the oil companies to sell regular diesel," he said.

But Burris has faith in his environmental theory of economics.

"If enough people demand it, someone will supply it," Burris said.

Oil companies aside, Burris' also has the fe 'eral government as an obstacle. While the Department of Energy is doing everything possible to promote alternative fuels, Congress is doing everything possible to make these fuels prohibitively expensive.

"Biofuel would not be much more expensive than regular diesel if it wasn't taxed," he said. "After we're done testing it, I'm going up to Washington to demand a tax credit to bring the price down. Right now, Biofuel is about 15 cents more a gallon than regular diesel. For a commercial fisherman using 10,000 gallons a month, that's too much."

Burris said government does not follow it's own environmental laws. "The EPA [Environmental Protection Agency] says government has to use the best available products for the environment," Burris said. "If they drag their feet on something so simple, they're not going to save this resource."

Although the National Oceanic and Athmospheric Administration endorses the fuel and uses it on their research vessels, the words "environmentally friendly fuel" do not appear anywhere in the Florida Keys National Marine Sanctuary plan.

"I think Billy Causey should be here and invite us to the table," Burris said. "What they want to do is going to take months and years. Here's something they can do that will immediately improve water quality."

According to Burris, using Biofuel has other benefits. "We could thumb our noses at the oil companies," he said. "And it would reduce the trade deficit. We make it here — right here in Florida."

Burris said he has no financial investments in Biofuel and no interest other than seeing the water improve. "We're killing the reef," he said. "And we're running out of time. In five years, the reef will be dead. It may die anyway. But trying to fix it now will clear our conscience a little."

Call Burris at (800) 634-5650 for information on where to obtain Biofuel.

Corpus Christi Caller

LIVING ■ 6B
ANN LANDERS ■ 7B

LOCAL NEWS B

Wednesday, October 10, 1984

Author Benchley, ex-shrimper campaign to save sea

GEORGE GONGORA/STAFF PHOTOGRAPHER

Author Peter Benchley

By Beth Arburn Davis
STAFF WRITER

Gary Burris believes the greatest danger facing mankind is not nuclear war, but decimation of the sea.

"Nuclear war is probable. This is certain," he said of the double jeopardies of pollution and over-fishing.

And although not all marine scientists agree, the former shrimp boat captain from Aransas Pass predicts the "point of no return" could be reached in as few as 15 years.

Burris is in Corpus Christi and South Texas to make a documentary about problems facing the oceans. Called "Sea Peace," the two-part, made-for-TV film will be narrated by award-winning author and screenwriter Peter Benchley of "Jaws," and is due for completion next spring.

Money for the $100,000 film has come from donations.

Burris, 34, is founder of a non-profit organization dedicated to awakening the public to the need for saving the sea. Survival of the Sea Society was founded approximately seven years ago and is based in Hot Springs, Ark. Burris said an office of his organization will be opened soon in Port Aransas.

During an interview yesterday at a Padre Island hotel serving as headquarters for his film crew, Burris said the oceans are dying from over-fishing and indiscriminate dumping of wastes. He said it is imperative that measures be taken to curb sport and commercial fishing, to establish fisheries that will replenish diminishing sealife, and to regulate commercial fishing boats to prevent inexperienced captains and crews.

Benchley, whose interest in the sea fueled his novels "Jaws," "The Deep," and "The Girl from the Sea of Cortez," said he has seen several instances where careless fishing techniques and pollution have produced underwater deserts.

"There are no more fish in Bermuda," he said, adding that the island in the Atlantic is an isolated microcosm that cannot be restocked because there are no nearby reefs. "The Bermuda rock fish, a delicacy, is gone."

He cited another instance in the Bahamas.

"I have seen Bahamian fishermen pour Clorox in reefs to get lobsters out. The reef never recovers. You can dynamite a reef and it will come back. But you pour Clorox on it and it doesn't come back."

Burris' statement that the ocean is quickly reaching a point of no return is not necessarily shared.

University of Texas marine scientist Dr. John Cullen said that there is disparity in the views of scientists regarding the problems of the oceans.

"It is my personal opinion that it's unquestionable that fishing has modified the environment, but in most marine areas there's not strong evidence that humans have affected production. In limited areas, however, there are serious consequences of human influence," he said, citing the Houston ship channel, the Hudson River plume and the Los Angeles sea area.

Cullen, whose specialty is plankton, the "food source" of the sea, said that while he is "hesitant to say categorically that the ocean is on the brink of death, we can't use the ocean indefinitely as a dumping ground."

GEORGE GONGORA/STAFF PHOTOGRAPHER

Filmmaker Gary Burris

The other way

Man heading to island of turmoil with Cuban aid

By ERIC DETWILER
Citizen Staff Writer

STOCK ISLAND — Though the number of refugees fleeing Cuba increases dramatically, one local man left Stock Island Sunday to take humanitarian aid to that impoverished island nation.

Before leaving, Gary Burris, owner and operator of the 50-foot vessel Dirk Seaguard, wondered aloud on whether he'll be allowed to return to the United States.

"I'm going out with the warning that they may close the ports," Burris said before sailing off. "I won't be let back into my home country."

Burris also says there's a chance that some Cuban citizens may try to hijack his boat and return to the U.S. He said Sunday that he's been told if that happens, he will be charged with smuggling.

During the last year and a half, Burris has made eight trips to Cuba carrying clothes, food, medicine — and occasionally ferrying journalists back and fourth.

On this trip, Burris and three crew members plan to take four tons of supplies — which they say will distribute to hospitals, children's homes and to needy families.

With the upheaval in Cuba over the last two weeks, Burris said he's not sure what kind of a reception he'll have when he arrives in Havana Tuesday.

"It's possible there will be trouble," Burris said. "I know one guy who was told to leave. I think the only reason they

See OTHER, Back page

J.D. DOOLEY / The Citizen

Capt. Gary Burris displays the load his vessel, the Seagurard, was to carry to Cuba beginning Sunday.

Emergency groups meet today

By ERIC DETWILER
Citizen Staff Writer

KEY WEST — The increasing numbers of Cuban refugees arriving in the Keys has spurred a meeting of state emergency planners on Stock Island today.

More than a hundred Cuban refugees arrived in the Keys Sunday — including two boats full of rafters that washed up on local beaches in the pre-dawn hours.

Coast Guard Group Key West reported a total of 85 refugees recovered in 10 separate cases. The Key West Police Department said they gathered up another 25 refuges on local beaches and took them to the Coast Guard Base.

See MEET, Back page

OTHER

From Page 1

would turn me away was if Havana was in danger of being overrun."

Burris says the Cuban government has given him free reign on where to distribute his goods on previous trips.

"Yes, there's probably things being skimmed from the top," Burris admitted, adding that most of the time he freely gives food to people who ask him for it.

"The Cubans have become comfortable enough with me to the point where it feels free," Burris said. "I don't try to do anything sneaky."

The humanitarian trips to Cuba have not endeared him to the South Florida Cuban community, Burris says. He claims the Cuban revolutionary group Alpha-66 has labeled him a "military objective."

"I'm not a Communist," Burris said. "I've had a lot of death threats."

Burris said if he's not allowed to come back to the Keys next week, he plans to head for the Bahamas or Mexico aboard the Seaguard.

J.D. DOOLEY / The Citizen

A man looks at a grounded Cuban refugee boat on South Roosevelt Boulevard Sunday. It was another busy day along the South Florida coast Sunday as more and more refugee boats were discovered and rafters rescued.

Captain Gary Burris

11661 Labrador Lane
East Naples, Florida 34140
(941) 272-0156

OBJECTIVE:

"Action through Awareness, Awareness through Education" as a way to help build a better world, and a more hopeful future for our children and grandchildren through environmental education.

QUALIFICATIONS

An accomplished environmental educator and documentary producer/writer with experience in front of, and behind, the camera. Well known as an effective environmental communicator capable of conveying environmental issues to the forefront of mass media and to the public's attention.

Through persistent efforts I have been instrumental in initiating and, or producing several awarding winning environmental/educational print and television programs. To accomplish many projects I have had working relations with CNN, Nickelodeon, PBS, The State of Rhode Island, The US Senate and House, the Cuban and Haitian Governments, Monroe County and Collier County Government.

In my educational work I have been responsible for the creation, production and publishing of an environmental program that included a magazine for schools, television documentaries and a live show for young people called the "ECOSHOW". This program was used as the basis of a Rhode Island model environmental program. I also initiated and co-wrote, with Anna Prager RI's Environmental Director under Governor Dupree, Rhode Island's model environmental education Bill. This was adopted by the Senate and passed in the 1st secession of the 101st Congress as S.1076

In my work I have also written and produced environmental themed songs for performance and public radio.

I can also captain a ship, fly a plane, birth a baby, cook a meal and plant a field etc.

EDUCATION

1969 Graduated Little Rock Central High
1969-72 State College of Arkansas

EMPLOYMENT

1965-1968 Little Rock Furniture Company:

This educational effort gained national attention as the model for the nation. President Bush sent his Secretary of Interior, Don Hodel, to endorse and kick off the project with great fanfare. In 1992 I began working in Cuba developing a "Green Model Project". Our Cuban environmental efforts resulting in the building of a biodiesel facility outside of Havana, and a facility to compost organic waste for soil reclamation.

2001-Founded Public Eyes-Public Ears Educational Radio and Television Company.

Since founding we have produced several timely documentaries on the condition of our environment and on solutions to our waste problems facing our country. In 2002 we began producing "Centerpoint Radio", an environmental issues radio talk show

REFERENCES

Peter Benchley, author
Jimmy Buffet, entertainer
Ted Kavanau retired founding VP CNN
Karen Childress, WCI
John Nocera, Naples, Florida

GREEN MODEL
DEVELOPMENT SEQUENCE
1981-1999

The development of the *Green Model* concept began in Arkansas, in 1981 with the first "Model State Campaign". At that time Survival of the Sea Society was founded and began gathering information and developing the environmental restoration/sustainable economy concept of an "Environmental Model/"

Since that time we have carried the development of this concept forward through research, education and entertainment. The focus that has developed in the last 17 years is one that, we believe will help implement a shift in awareness of alternatives and of product awareness.

The central focus today is to establish and publicize the relationship between possible environmental benefit and true cost savings brought by making the change to alternative technology and products. This understanding will help individuals, and municipal and foreign governments decide to take steps to change their methods of environmental management by incorporating the "*Green Model*" into policy.

This awareness shift is now occurring in the Southwest Florida and in the Florida Keys through education, media, and be developing a county - city *Green Model* with local, state, federal agencies involved where possible to the benefit of all involved.

"Think globally and Act locally" is the operating philosophy of building the "Model."

The first "*Green Model*" will be a "Greenprint" (blueprint) for sustainable change. That *Greenprint* will utilize all the best-proven technology, available or developing, worldwide, to reduce a community's or a country's pollution loading and unburdens the existing system: An exportable *Greenprint* when completed.

To best describe the sequence of this project I must continue with a history of our organization, which led to the current Green Model/Biofuel Project.

"Survival of the Sea Society" was founded and began an information gathering effort in 1981. The objective was to gather all information regarding the condition of the ocean including the effects of river pollution, thus ascertaining a real time condition of the environment. Thinking that if people really understood then they would act on that information and make changes.

The results of this campaign were disseminated through all the available media and used to develop teaching material. The efforts of this state-based project gained some national attention via television reports, talk shows, radio shows, and University lecture tours throughout the country.

The stage of development reached in Arkansas was first called the "Model State" program and focused a great deal on rural solutions to water and agribusiness pollution of the river system of the state.

By 1984 we had developed original environmental documentary evidence of the rapid decline of the coastal United States, and its river systems. We formed our own television group and began work on the project.

We enlisted and received the support and help from both Jimmy Buffet and Jaws author, Peter Benchley. At that time Benchley was on the best seller's list and a very big draw. His services as writer and narrator on National Geographic films at the time were highly acclaimed. He joined the board of Survival of Sea Society and remains involved today.

In the fall of 1984 we teamed up with the fledgling CNN, and utilized the researched material, and the new production group, to produce an award winning documentary and news series with Turner's group called the Coastline Crisis

This project ran through 1985 and led to our group producing an environmental series "The Ocean Report". By early 1987 we had been approached by the U.S. EPA to produce a series of television documentaries. The videos would cover estuaries and sanctuaries nationwide. The Congress had passed the National Sanctuary Act: Our job was to educate the public concerning the condition of coastal waters and the significance of those resources to the Nation.

Out of this project, which began in Rhode Island's Narragansett Bay and Long Island Sound and ended up in Puget Sound, we furthered the development of the *'Green Model'* concept.

In Rhode Island we worked with the idea that, it being the smallest state, we could develop a plan to 'fix all possible environmental problems and become the demonstrable example of environmental recovery for the Nation.' Working with the Rhode Island Governor and his environmental staff (Chief of staff, Anna Prager) we wrote one of the nation's first comprehensive mandatory environmental education bill for all schools. We produced the education material for this program and that material has been reproduced in several states. Our state bill was copied by U.S. Senator John Chaffe's staff and was passed as the National Environmental Education Bill by the United States Congress

This project produced another award winning documentary. After seeing the video, Never a Sabbath, in which he was the on camera protagonist for the environment, Peter Benchley publicly stated that it 'was the most comprehensive environmental video ever produced, or ever will be, on the subject it covered.'

In 1988, carried by the first success of the Rhode Island Project, we continued with the next phase of the environmental material development. We were funded further by a number of governmental agencies and individuals to create the 'Model's' teaching tools and demonstrate concepts for teachers to teach the environment on a scale of understanding that was global, but, dealt with the local issues. This state campaign was called the "Better Bay through Better Education."

This educational effort gained national attention as the model for the nation. President Bush sent his Secretary of Interior, Don Hodel, to endorse and kick off the project with great fanfare. A number of Governors from the surrounding states attending at the request of Rhode Islands' Governor Edward Deprie.

The educational material resulting from that work has been, and is being, used in all of RI public schools. The work has been duplicated in a number of states. That work featured an environmental tabloid that encompassed all the 'vital understanding' that we felt the youth and adults of the state and the world needed. A Rhode Island Board of teachers approved it for all grades.

By 1989 we began assembling the various educational components and environmental products we had learned about into what we called the 'World's First EcoShow'.

in front of hundreds of reporters in a Havana press conference, the idea of developing Cuba into a "Global Green Model". The press seized the idea and reported it worldwide, but especially so in the Spanish press and television. Many of the ideas presented to the Cuban Environmental Ministry have been utilized.

After that conference, on a later trip delivering supplies and environmental education materials, I was invited to speak on Radio Rebeldi or Rebel Radio, to the nation, on the subject of becoming a 'Green Model'. Resulting from that work an opportunity existed to develop alternative waste treatment projects in Cuba that was instrumental in cleaning the upstream water and air which moves over and past Florida.

A major step in this 'Model" has been accomplished by a cooperative effort between, the City of Naples, Clean Cities and the Seaguard's *Green Model Project*. A full circle environmental demonstration is nearing completion and implementation. This will be the true beginning of a long developing; world-changing community supported municipal/environmental model for the nation.

The first show started humbly on the City dock in New London, CT., with two entertainers and a few environmental education displays. We also acquired our 'Flagship', an 80-foot former North Sea iron sailing fishing ship. We began touring coastal events and waterfront festivals that year. Our show traveled from May until September from Boston to Norfolk, VA.

During 1989 we had founded the Seaguard with the idea in mind of establishing oil spill response squads to be ready for spills along the coast. To help equip such squads we enlisted the help of Congresswoman Claudine Sneider. Out of that collaboration we were able to get the congress to include that idea in the Oil Spill Liability Act of placing spill response equipment along the America. That act and our idea of response squads have help keep America's coast a little cleaner.

The following year we wrote and produced an album of Eco-songs and expanded the show to over twenty performers. This show grew to tour extensively during the next two years. Our last major show drew over twenty thousand children and their parents during a two-day event in Providence Rhode Island. (See enclosed letters of support)

During 1991 and 1992 we continued developing educational material and videos footage for use in a planned future television series on the solutions to some of the oceans and the planets environmental problems.

By 1992, to expand the scope of our television research, we moved our efforts and the Seaguard, our office/studio/ship from the New England area. We toured to the south from Maryland to the Florida to have a look at the conditions of other fisheries and histories of those areas. We showed our newest video on many local PBS stations at the ports we stopped in and we appeared on many radio, and television shows. Our documentary work along the East Coast focused a lot of attention on the Carolinas.

By early summer of 1992, we had joined forces with an environmental television group out of England and Florida. That project took us to the Bahamas and a lengthy production with children from around the world and wild dolphins in nature who willingly interact with humans. This work is currently airing in England, Europe and all Spanish speaking countries including Cuba. As a result of this project in the Bahamas we have established a working relationship with an environmental education group in the Bahamas. They will be partners in future recycling projects. We provided them with songs and other education material that is currently in use in that country.

Late summer of '92 found us back in the country, and in Miami, in the eye of Hurricane Andrew. Our ship sustained damage. We recovered after several months of work on the ship and relocated our headquarters to Key West, Florida in the fall.

During the next few months we increased our efforts in gathering environment product and technology information. Many new products are slowing emerging into practical use and we have followed those developments. As new products or technology are brought to the marketplace or can be use in a working situation we add this product/technology to our emerging *'Green Model'* development.

By the spring of 1993 we expanded the scope of the 'Green Model' to the level establishing a Model Island Project. The opportunity to attempt this came in April of 1993 when we joined a humanitarian flotilla to Cuba.

This successful trip was extremely controversial, and well covered by the media (0ver 300 reporters and 90 satellite dishes). We had the opportunity to propose to the Cuban Officials,

<u>Millennium(Green) Model</u>

A Seaguard/Survival of the Sea Society Project

Vision for the Business of the Millennium

The business vision of <u>Millennium (Green) Model</u>, is built on the knowledge that we have damaged our planet environmentally nearly to the point of no-return.

After two decades of educational activity, on many environmental and humanitarian fronts, we still almost daily witness or hear of our Earth's continued decline.

My organization, The Seaguard/Survival of the Sea Society, had to take a new direction so as not to work in vain.

Seeing that little has changed and, something must happen, our direction had to become even more pragmatic in our activism for the environment. We understand that "sound business" is the quickest pathway to environmental recovery. It is the best way that we can make change happen more quickly than government.

Understanding the reality of our Planet's environmental situation has given us <u>no option</u> but to take "business" steps to stop as much of the damage as possible. A move into "pragmatic environmental business activism" is required to accomplish any substantial improvements in the Earth's environment.

How do we intend to do this? We intend to do this through implementing a causal shift in product awareness, and we accomplish that task through popular educational entertainment , agressive use of the media worldwide, and by developing a <u>Green Model</u>: a profitable Model.

The first "Green Model" will be a blueprint for sustainable change. That blueprint will utilizes all the best proven technology available or developing, worldwide, to reduce a community's or a country's pollution loading and unburden the existing system: An exportable blueprint.

The business problem that we must solve is to develop a way to accomplish this profitably so other communities will follow suit without hesitation. .

There are two things that must happen at the same time to make this effort succeed and count substantially: Education and workable affordable solutions. If these two forces are combined into a business ethic we will prosper.

There is big business in repairing the Earth.

This we should fear. Poison in our food

gathering grounds. Drain the sea and it would look like Sanford's backyard.

This on my deck reality fueled my activist starting point

www.ingramcontent.com/pod-product-compliance
Lightning Source LLC
Chambersburg PA
CBHW060819170526
45158CB00001B/21
* 9 7 8 1 0 8 2 4 7 2 4 7 3 *